Lecture Notes in Mechanical Engineering

Lecture Notes in Mechanical Engineering (LNME) publishes the latest developments in Mechanical Engineering—quickly, informally and with high quality. Original research reported in proceedings and post-proceedings represents the core of LNME. Volumes published in LNME embrace all aspects, subfields and new challenges of mechanical engineering. Topics in the series include:

- Engineering Design
- Machinery and Machine Elements
- Mechanical Structures and Stress Analysis
- Automotive Engineering
- Engine Technology
- Aerospace Technology and Astronautics
- Nanotechnology and Microengineering
- Control, Robotics, Mechatronics
- MEMS
- Theoretical and Applied Mechanics
- Dynamical Systems, Control
- Fluid Mechanics
- Engineering Thermodynamics, Heat and Mass Transfer
- Manufacturing
- Precision Engineering, Instrumentation, Measurement
- Materials Engineering
- Tribology and Surface Technology

To submit a proposal or request further information, please contact the Springer Editor of your location:

China: Ms. Ella Zhang at ella.zhang@springer.com
India: Priya Vyas at priya.vyas@springer.com
Rest of Asia, Australia, New Zealand: Swati Meherishi at swati.meherishi@springer.com
All other countries: Dr. Leontina Di Cecco at Leontina.dicecco@springer.com

To submit a proposal for a monograph, please check our Springer Tracts in Mechanical Engineering at http://www.springer.com/series/11693 or contact Leontina.dicecco@springer.com

Indexed by SCOPUS. All books published in the series are submitted for consideration in Web of Science.

More information about this series at http://www.springer.com/series/11236

L. M. Das · Abhishek Sharma · Fitwi Yohaness Hagos · Sumit Tiwari
Editors

Recent Trends in Thermal Engineering

Select Proceedings of ICCEMME 2021

Springer

Editors
L. M. Das
Department of Mechanical Engineering
Centre for Energy Studies
IIT Delhi
New Delhi, Delhi, India

Abhishek Sharma
Department of Mechanical Engineering
G. L. Bajaj Institute of Technology
and Management
Greater Noida, Uttar Pradesh, India

Fitwi Yohaness Hagos
Department of Mechanical
and Industrial Engineering
Sultan Qaboos University
Muscat, Oman

Sumit Tiwari
Department of Mechanical Engineering
School of Engineering
Shivnadar University
Greater Noida, Uttar Pradesh, India

ISSN 2195-4356 ISSN 2195-4364 (electronic)
Lecture Notes in Mechanical Engineering
ISBN 978-981-16-3427-7 ISBN 978-981-16-3428-4 (eBook)
https://doi.org/10.1007/978-981-16-3428-4

This Springer imprint is published by the registered company Springer Nature Singapore Pte Ltd.
The registered company address is: 152 Beach Road, #21-01/04 Gateway East, Singapore 189721, Singapore

Preface

International Conference on Computational and Experimental Methods in Mechanical Engineering (**ICCEMME 2021**) has been the third conference of its series organized by the Department of Mechanical Engineering of G. L. Bajaj Institute of Technology and Management, Greater Noida, Uttar Pradesh, India. The institute is located in the vicinity of the industrial hub. Therefore, it was decided to provide a forum to bring together scientists, speakers from industries, university professors, graduate students, and mechanical engineers, presenting new research in science, technology, and engineering.

This book includes Research articles from various areas of thermodynamics, heat transfer, computational fluids dynamics, multiphase fluids fluid measurement, as well as thermal energy systems including internal combustion engines, refrigeration, and heat exchangers. The group members have experience of carrying out cutting-edge and international-recognized researches. The other members' expertise matches and can strongly support the above research areas. The cross-disciplinary nature of fluids and thermal engineering, interactive expertise in fluids, thermodynamics, two-phase flow and phase changes, and heat transfer can help promote the individual's collaborations in not only fundamental researches but also applied technological studies. During the conference, about ten delegates were joined from various countries and delivered keynote lectures on the theme of the conference. All papers were critically reviewed by two reviewers from national/international authors.

Furthermore, we would like to extend our appreciation to all the authors for contributing their valuable research in the conference. The committee is also grateful to all reviewers who spared their time to carefully review all the assigned research articles and to all the committee members for their great effort for making the conference a grand success.

We are thankful to all sponsored agencies who gave us their cooperation and funding support.

We are thankful to our management and director of G. L. Bajaj Institute of Technology and Management, Greater Noida, Uttar Pradesh, India, for their continuous source of inspiration and valuable support. We are thankful to all the members of the organizing committee for their contribution in organizing the conference. Last but not least, we thank Springer for its professional assistance and particularly Ms. Priya Vyas and Ms. Sushmitha Shanmuga Sundaram who supported this publication.

New Delhi, India — Prof. (Dr.) L. M. Das
Greater Noida, India — Prof. (Dr.) Abhishek Sharma
Muscat, Oman — Dr. Fitwi Yohaness Hagos
Greater Noida, India — Dr. Sumit Tiwari

Contents

Editors and Contributors

About the Editors

Dr. L. M. Das is a retired Professor from IIT Delhi. He holds a Ph.D. degree in Mechanical Engineering from IIT, Delhi. Dr. Das has got a teaching/research experience of more than forty four years to his credit. His areas of research interest centres around development of alternative-fuelled low emission engines/vehicles. He has supervised 34 Ph.D. theses and seventy-seven M.Tech. Projects. He has published more than 80 research papers in various international/national journals. He had several international research assignments. He was a Visiting Professor in University of California, Riverside (Centre for Environment Research and Technology) in USA; researcher in INRETS, France; Consultant to Shell (Thornton Research Centre, Chester, England); and researcher in Lisbon Technical University (Portugal). He had delivered invited lectures in India as well as in USA, Canada, England, France, Germany, Japan, Switzerland, Australia, Belgium Netherlands, Thailand, Moscow, Turkey, Philippines, Nepal and Sri Lanka. He has been a member of several research/technical and scientific advisory committees of Department of Science and Technology, Ministry of New and Renewable Energy, Indian Institute of Petroleum, Indian Council of Agricultural Research, Department of Biotechnology, Indian Oil Research Lab, etc. He was a Member of Core group of Automotive Research, (CAR) and a member of the Expert Committee for Auto-Fuel Vision and Policy 2025. He was the Convener of Technical Committee of World Hydrogen Technology Convention held in August 2009 at New Delhi. Presently he is also the Chairman of Sectional Committee of PCD 3 of Bureau of Indian Standards. Professor Das had been the principal investigator of several sponsored projects funded by MNRE, DST, General Motors (USA), European Union, UNIDO, Shell (England). He has been awarded "Rajiv Gandhi Samman" by the Department of Science and Technology, Government of Orissa (2005). In 2012, he has been awarded Lockheed Martin award for his R&D achievements. International Association of Hydrogen Energy has given him the "Outstanding Service Award" for his "Altruistic Contribution" to the "cause of Hydrogen

Economy". He has also been awarded T. N. Veziroglu award by Hydrogen Association of India (2013) and Biju Patnaik award by Odisha Science Academy for scientific excellence "in recognition of his original and creative contribution" (2014).

Dr. Abhishek Sharma working as a Professor in Department of Mechanical Engineering, G. L. Bajaj Institute of Technology and Management, Greater Noida, India. He has more than 17 years of international involvement in academic teaching and research across recognized Indian and African universities. He has published over fifty research articles in SCI/SCIE and SCOPUS indexed journals; other peer-reviewed journals; books; IEEE conference proceedings. He serves as the reviewer and editorial member in many peer-reviewed journals and conferences. He received his bachelor degree in Mechanical Engineering from HBTI Kanpur, India, in 2003; Master of Technology (M.Tech.) in Thermal Engineering from Uttar Pradesh Technical University, Uttar Pradesh, India, in 2011; Doctor of Philosophy (Ph.D.) in Mechanical Engineering from UPES Dehradun, India, in June 2016. He is associated with academic teaching and research by functioning as a research committee member; master thesis examiner; university curriculum revision committee member; course instructor; university question setter; master program evaluator; journal editor; SCIE and Scopus-indexed journals referee; book author; and IEEE conference referee. His primary areas of research interest include the development of alternative-fueled low emission engine/vehicles, alternative fuels such as CNG, biodiesel and their blends and environmental conscious engine modeling.

Dr. Fitwi Yohaness Hagos is currently working as a senior lecturer in the Department of Mechanical and Industrial Engineering, College of Engineering, Sultan Qaboos University, Al Khould, Muscat, Oman. He also worked as a senior lecturer in the Universiti Malaysia Pahang, Malaysia, since February 2014 to November 2020. He received his Ph.D. degree in Mechanical Engineering, Universiti Teknologi PETRONAS, and Malaysia in 2014. He received his Master of Engineering from Delhi College of Engineering, Delhi University, India in 2005. He received his B.Sc. Degree in Mechanical Engineering from Addis Ababa University, Ethiopia in 2001. He has served from graduate assistant until the rank of Assistant Professor from 2002 to 2010 in the Mechanical Engineering Department, Mekelle University. During his service in Mekelle University, he has assumed different academic and administrative posts. He has published more than 60 articles in international scholarly journals and conferences. He has also supervised 1 Ph.D. and 1 Masters by research thesis, 7 Master thesis (mixed mode), 1 Masters by coursework dissertation and more than 40 undergraduate final year projects. Currently he has 1 Ph.D. and 3 Masters Students under his supervision. His area of specialization is IC engines, thermo-fluids and energy and his research interests are in IC engines, combustion, modelling and simulation, optical diagnostics applied to combustion and flow, thermo-fluids, energy storage, renewable energy, energy resource assessment and automotive engineering. He is certified chartered Engineer by the Engineering council, UK, and Member of Institute of Mechanical Engineers (IMechE), UK. Besides, he is also an editorial board member for *Biofuels* Journal and *Brana Journal of Engineering*. Since January 2015, he is associate editor-in-chief for the *Journal of Mechanical Engineering and Science* (JMES).

Dr. Sumit Tiwari is working with Shiv Nadar University, Dadri, Uttar Pradesh. He received the B.Tech. degree in mechanical engineering from the Uttar Pradesh Technical University, Lucknow (now AKTU). Further, he got M.Tech. and Ph.D. degree from the Indian Institute of Technology (IIT) Delhi. He got assistantship for his Ph.D. from "Ministry of New and Renewable Energy (MNRE)" assistantship. He worked with Professor Patrick Phelan's Group, USA, as a research intern for 06 month under the fellowship named "Building Energy Efficiency Higher and Advanced Network (BHAVAN) Fellowship Program" which is a program executed by Indo-U.S. Science and Technology Forum (IUSSTF) and department of science and technology (DST). He got Shrimati Vijay-Usha Sodha Research Award for the session 2018–19 given by Center for Energy Studies, IIT Delhi. He also got second prize under "Best thesis award" in Ph.D. symposium organised by "Madan Mohan Malaviya University of Technology, Gorakhpur, Uttar Pradesh, India and University of the Ryukyus, Okinawa, Japan". He is working on different solar thermal technologies namely, photovoltaic thermal (PVT) greenhouse drying system, PVT greenhouse heating system, PVT greenhouse biogas heating system, solar still, PVT air collector, PVT air collector integrated drying system, passive cooling of building, and solar adsorption cooling system etc. He has published several SCI journal papers with high impact factor and international conference papers in his field. He has more than 12 years' experience of teaching and research. He has also got several recognitions during his study as well as in teaching. He is working as a reviewer in more than 17 International SCI Journals.

Contributors

Wasim Akram Mechanical Engineering Department, National Institute of Technology Jamshedpur, Jamshedpur, India

Amardeep Department of Mechanical Engineering, GL Bajaj Institute of Technology and Management, Greater Noida, India

Md Yeashir Arafat Department of Mechanical Engineering, Bangladesh University of Engineering and Technology, Dhaka, Bangladesh

Navid Aslfattahi Department of Mechanical Engineering, University of Malaya, Kuala Lampur, Malaysia

Harshit Bahri Department of Mechanical Engineering, G L Bajaj Institute of Technology and Management, Greater Noida, Uttar Pradesh, India

Suraj Bhan Department of Mechanical Engineering, Delhi Technological University, Delhi, India

Shashwata Chakraborty Joint School of Nanoscience and Nanoengineering, North Carolina A&T State University, Greensboro, NC, USA

Nathi Ram Chauhan Department of Mechanical and Automation Engineering, Indira Gandhi Delhi Technical University for Women, New Delhi, India

Pankaj Kumar Chauhan Department of Mechanical Engineering, G.L. Bajaj Institute of Technology and Management, Greater Noida, India

Abhay Dahiya Department of Mechanical Engineering, G.L Bajaj Institute of Technology and Management, Greater Noida, India

Debashis Datta Nuclear Safety and Security Division, Bangladesh Atomic Energy Regulatory Authority, Dhaka, Bangladesh

Akash Deep Department of Mechanical Engineering, G.L. Bajaj Institute of Technology and Management, Greater Noida, India

Raghvendra Gautam Department of Mechanical Engineering, Delhi Technological University, Delhi, India

Tarun Kumar Gupta Department of Mechanical Engineering, G.L. Bajaj Institute of Technology and Management, Greater Noida, India

Udit Gupta B.Tech. Student, Department of Mechanical Engineering, G.L. Bajaj Institute of Technology and Management, Greater Noida, Uttar Pradesh, India

M. A. Hassan Mechanical Engineering Department, National Institute of Technology Jamshedpur, Jamshedpur, India

Ritesh Jaiswal Department of Mechanical Engineering, Kamla Nehru Institute of Technology, Sultanpur, Uttar Pradesh, India

Manav Jha Department of Mechanical Engineering, GL Bajaj Institute of Technology and Management, Greater Noida, India

Junaid Khan Sustainable Energy and Acoustics Research Laboratory, Department of Mechanical Engineering, Aligarh Muslim University, Aligarh, India

Mohammad Shahbaz Khan Department of Mechanical Engineering, G.L. Bajaj Institute of Technology and Management, Greater Noida, Uttar Pradesh, India

Ambuj Kumar Department of Mechanical Engineering, Kamla Nehru Institute of Technology, Sultanpur, Uttar Pradesh, India

Amit Kumar Department of Mechanical Engineering, National Institute of Technology, Srinagar (Garhwal), Uttarakhand, India

Devendra Kumar Department of Mechanical Engineering, Sachdeva Institute of Technology, Farah, Mathura, Uttar Pradesh, India

Manoj Kumar G. B. Pant University of Agriculture and Technology, Pantnagar, Uttrakhand, India

Uday Kumar B.Tech. Student, Department of Mechanical Engineering, G.L. Bajaj Institute of Technology and Management, Greater Noida, Uttar Pradesh, India

Mahesh Kumar Maddheshiya Department of Mechanical Engineering, GL Bajaj Institute of Technology and Management, Greater Noida, India

Naman Parashar Sustainable Energy and Acoustics Research Laboratory, Department of Mechanical Engineering, Aligarh Muslim University, Aligarh, India

Lalta Prasad Department of Mechanical Engineering, National Institute of Technology, Srinagar (Garhwal), Uttarakhand, India

Roshan Raman Assistant Professor, Department of Mechanical Engineering, The NorthCap University, Gurugram, India

Pooja Rani G L Bajaj Institute of Technology and Management, Greater Noida, Uttar Pradesh, India

Prabhat Ranjan Department of Mechanical Engineering, G.L. Bajaj Institute of Technology and Management, Greater Noida, India

Souvik Singh Rathore Mechanical Engineering Department, National Institute of Technology Jamshedpur, Jamshedpur, India

R. Saidur School of Science and Technology, Sunway University, Selangor, Malaysia

Ravi Kumar Samadhiya Department of Mechanical Engineering, Sachdeva Institute of Technology, Farah, Mathura, Uttar Pradesh, India

Sanjay Sanjay Mechanical Engineering Department, National Institute of Technology Jamshedpur, Jamshedpur, India

Manish Saraswat Department of Mechanical and Automation Engineering, Indira Gandhi Delhi Technical University for Women, New Delhi, India;
Department of Mechanical Engineering, Lloyd Institute of Engineering and Technology, Greater Noida, India

Rahul Shakye B.Tech. Student, Department of Mechanical Engineering, G.L. Bajaj Institute of Technology and Management, Greater Noida, Uttar Pradesh, India

Abhishek Sharma Department of Mechanical Engineering, G L Bajaj Institute of Technology and Management, Greater Noida, Uttar Pradesh, India

Payal Sharma Department of Mechanical and Automation Engineering, Indira Gandhi Delhi Technical University for Women, New Delhi, India

Rahul Sharma Assistant Professor, Department of Mechanical Engineering, G.L. Bajaj Institute of Technology and Management, Greater Noida, Uttar Pradesh, India

Sagar Sharma Department of Mechanical Engineering, GL Bajaj Institute of Technology and Management, Greater Noida, India

Vineet Kumar Sharma Department of Applied Sciences, G. L. Bajaj Institute of Technology and Management, Greater Noida, India

Anoop Singh Department of Mechanical Engineering, GL Bajaj Institute of Technology and Management, Greater Noida, India

Devendra Singh Department of Mechanical Engineering, Sachdeva Institute of Technology, Farah, Mathura, India

Harvendra Singh Department of Mechanical Engineering, G L Bajaj Institute of Technology and Management, Greater Noida, Uttar Pradesh, India

Kaushalendra Kumar Singh Department of Mechanical Engineering, G L Bajaj Institute of Technology and Management, Greater Noida, Uttar Pradesh, India

Padam Singh Department of Applied Sciences, Galgotias College of Engineering and Technology, Greater Noida, India

Prajjwal Singh Department of Mechanical Engineering, GL Bajaj Institute of Technology and Management, Greater Noida, India

Pusphendra Singh Department of Mechanical Engineering, Delhi Technological University, Delhi, India

Ranjeet Kumar Singh Department of Mechanical Engineering, G.L. Bajaj Institute of Technology and Management, Greater Noida, India

Vimal Pratap Singh B.Tech. Student, Department of Mechanical Engineering, G.L. Bajaj Institute of Technology and Management, Greater Noida, Uttar Pradesh, India

Himanshu Tanwar B.Tech. Student, Department of Mechanical Engineering, G.L. Bajaj Institute of Technology and Management, Greater Noida, Uttar Pradesh, India

Omprakash Yadav Department of Mechanical Engineering, The NorthCap University, Gurugram, India;
Assistant Professor, Department of Mechanical Engineering, G.L. Bajaj Institute of Technology and Management, Greater Noida, Uttar Pradesh, India

Vinod Kumar Yadav Department of Mechanical Engineering, G.L. Bajaj Institute of Technology and Management, Greater Noida, Uttar Pradesh, India

Virendra Kumar Yadav Department of Mechanical Engineering, Kamla Nehru Institute of Technology, Sultanpur, Uttar Pradesh, India

Syed Mohd Yahya Sustainable Energy and Acoustics Research Laboratory, Department of Mechanical Engineering, Aligarh Muslim University, Aligarh, India

Transport Phenomena in a PWR Subchannel Replete with Al_2O_3–TiO_2/Water Hybrid Nanofluid: A CFD Approach

Md Yeashir Arafat, Shashwata Chakraborty, and Debashis Datta

Nomenclature

c_p	Specific heat
d	Diameter
f	Friction factor
g	Gravitational acceleration
h	Convective heat transfer coefficient
i, j	Unit vector
k	Turbulence kinetic energy
l	Length
Nu	Nusselt number
P	Pressure
q''	Heat flux
Re	Reynolds number
T	Temperature
u	Velocity component
u'	Turbulent fluctuating quantity
V	Velocity
x, y, z	Coordinates

M. Y. Arafat (✉)
Department of Mechanical Engineering, Bangladesh University of Engineering and Technology, Dhaka, Bangladesh

S. Chakraborty
Joint School of Nanoscience and Nanoengineering, North Carolina A&T State University, Greensboro, NC, USA
e-mail: schakraborty@aggies.ncat.edu

D. Datta
Nuclear Safety and Security Division, Bangladesh Atomic Energy Regulatory Authority, Dhaka, Bangladesh
e-mail: datta@baera.gov.bd

L. M. Das et al. (eds.), *Recent Trends in Thermal Engineering*,
Lecture Notes in Mechanical Engineering,
https://doi.org/10.1007/978-981-16-3428-4_1

Greek Letters

μ Dynamic viscosity
ρ Density
λ Thermal conductivity
φ Nanoparticle volume fraction

Subscripts

av Average
b Bulk
bf Base fluid
h Hydraulic
in Inlet
nf Nanofluid
np Nanoparticles
w Wall

1 Introduction

The concept of nanofluid introduced a couple of decades ago has thrived in numerous engineering applications including industrial cooling, nuclear reactors, automotive cooling, heat exchangers, biomedical field, and so on. In general, nanofluids are prepared by dispersing individual solid particles, fibers, or tubes with sizes less than 100 nm in various base fluids having thermal conductivity remarkably lower than that of solids. The commonly used nanoscale particles also called nanoparticles are Al_2O_3, CuO, TiO_2, SiO_2, Cu, SiC, etc., while H_2O and $C_2H_6O_2$ are the most traditional base fluids [1]. The advent of hybrid nanofluid, an emerging technique of suspending mixture or composite of dissimilar nanoparticles in a particular base fluid, has taken the nanotechnology-based thermofluids to another dimension. The potentiality of Al_2O_3–TiO_2/water hybrid nanofluid has grown significant contemplation among the researchers that are well reflected in present-day studies in the literature [2].

Energy efficiency is of prime concern in nuclear power generation which is substantially hindered owing to inadequate heat removal from the reactor core. Therefore, nanofluids are being carefully considered to enhance the cooling capacity as well as the safety aspects of a nuclear reactor [3]. Several researchers have made notable contributions till date to evaluate the possibility and compatibility of different nanofluids as reactor coolants. Nazififard et al. [4] observed a significant pressure drop in the PWR subchannel upon replacing pure water with

Al_2O_3/water nanofluid. Ghazanfari et al. [5] noticed a reduction in central fuel temperature, while alumina nanoparticles are dispersed in reactor coolant. Bahrevar et al. [6] identified a lower pressure drop in the SCWR subchannel due to the application of alumina nanofluid as a coolant. Saadati et al. [7] reported the nuclear reactor to be safer when Ag/H_2O nanofluid is chosen as coolant ascribed to high negative reactivity temperature coefficient.

Though a number of researches have been conducted involving alumina, titania, or zirconia nanofluids in nuclear reactor subchannel, authors are not aware of the implication of any hybrid nanofluid for reactor cooling. Therefore, in this article, efforts are aimed to numerically analyze the cooling performance of Al_2O_3–TiO_2/water hybrid nanofluid in a PWR subchannel under turbulent forced convection conditions. The nanoparticle concentration is varied from 1 to 4% while the Reynolds number from 20,000 to 80,000. The effectiveness of the hybrid nanofluid compared to its base fluid is evaluated based on heat transfer rate, frictional loss, and thermal performance factor as per the recommendation of IAEA Safety Reports Series No. 30 [8].

2 Thermophysical Properties

The homogeneous mixture of aluminum oxide and titanium oxide nanoparticles suspended in pure water is assumed to be a single-phase flow, isotropic, incompressible, and Newtonian fluid [9]. The velocity and temperature of dispersed hybrid nanoparticles are considered to be equal to those of base fluid [9].

The effective thermophysical properties $\left(\text{i.e.}, \rho_{\text{nf}}, (c_p)_{\text{nf}}, \lambda_{\text{nf}}, \mu_{\text{nf}}\right)$ of hybrid nanofluid can be determined from the following formulae of Sharma et al. [10].

$$\rho_{\text{nf}} = \left(\frac{\varphi}{100}\right)\rho_{\text{np}} + \left(1 - \frac{\varphi}{100}\right)\rho_{\text{bf}} \tag{1}$$

$$\left(\rho c_p\right)_{\text{nf}} = \frac{\varphi}{100}\left(\rho c_p\right)_{\text{np}} + \left(1 - \frac{\varphi}{100}\right)\left(\rho c_{\text{np}}\right)_{\text{bf}} \tag{2}$$

$$\lambda_{\text{nf}} = \lambda_{\text{bf}}\left[0.894\left(1 + \frac{\varphi}{100}\right)^{1.37}\left(1 + \frac{T_{\text{nf}}}{70}\right)^{0.278}\left(1 + \frac{d_{\text{np}}}{100}\right)^{-0.034}\left(\frac{\alpha_{\text{np}}}{\alpha_{\text{bf}}}\right)^{0.017}\right] \tag{3}$$

$$\mu_{\text{nf}} = \mu_{\text{bf}}\left[\left(1 + \frac{\varphi}{100}\right)^{11.3}\left(1 + \frac{T_{\text{nf}}}{70}\right)^{-0.038}\left(1 + \frac{d_{\text{np}}}{170}\right)^{-0.061}\right] \tag{4}$$

The thermophysical properties of base fluid (water) and solid nanoparticle (Al_2O_3, TiO_2) are specified in Table 1 [11]. Table 2 presents the effective

Table 1 Thermophysical properties of solid nanoparticles and base fluid [11]

Nanoparticles/base fluid	ρ (kg/m^3)	c_p (J/kg K)	λ (W/mK)	μ (Ns/m^2)
Al_2O_3	3,970	765	40	–
TiO_2	4,250	686.2	8.95	–
water	997.1	4,179	0.613	0.001

Table 2 Effective thermophysical properties of Al_2O_3–TiO_2/water hybrid nanofluid

φ	ρ_{nf} (kg/m^3)	$(c_p)_{nf}$ (J/kg K)	λ_{nf} (W/mK)	μ_{nf} (Ns/m^2)
1%	1,028.23	4,040.96	0.635	0.001097
2%	1,059.36	3,911.04	0.643	0.001227
4%	1,121.62	3,672.82	0.661	0.001528

thermophysical properties of the hybrid nanofluid calculated for the nanoparticle diameter of 30 mm, nanofluid temperature of 293 K, and nanofluid concentrations (φ) of 1, 2, and 4%.

3 Mathematical Modeling

The steady-state, turbulent-forced convection flow has been employed on the subchannel of a standard PWR occupied by Al_2O_3–TiO_2/water hybrid nanofluid. Figure 1 shows the schematic view and meshing surfaces of the subchannel considered in the numerical analysis. The design parameters of the subchannel model having flow symmetry are adopted from the previous study in the literature and presented in Table 3 [12]. The continuity, momentum, and energy equations which govern the mathematical model are as follows [12].

$$\frac{\mathrm{d}\langle u_i\rangle}{\mathrm{d}x_i} = 0 \tag{5}$$

$$\frac{\mathrm{d}\langle u_i\rangle\langle u_j\rangle}{\mathrm{d}x_j} = -\frac{\mathrm{d}\langle P\rangle}{\mathrm{d}x_i} + \frac{\mathrm{d}}{\mathrm{d}x_j}\left[\mu\left(\frac{\mathrm{d}\langle u_i\rangle}{\mathrm{d}x_j} + \frac{\mathrm{d}\langle u_j\rangle}{\mathrm{d}x_i}\right) - \rho\left\langle u_i' u_j'\right\rangle\right] + \rho g_i \tag{6}$$

$$\frac{\mathrm{d}}{\mathrm{d}x_i}(\langle u_i\rangle(\rho E + P)) = \frac{\mathrm{d}}{\mathrm{d}x_i}\left(k_{\mathrm{eff}}\frac{\mathrm{d}T}{\mathrm{d}x_i} + u_j(\tau_{ij})_{\mathrm{eff}}\right) \tag{7}$$

For simulating turbulence in ANSYS FLUENT, the k-ε turbulence model has been applied [9].

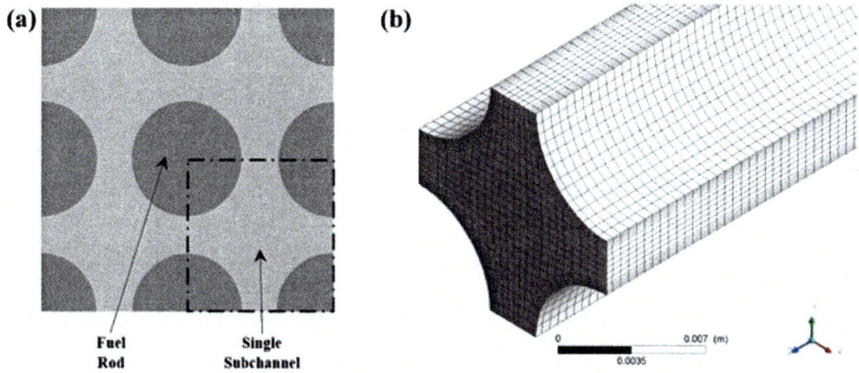

Fig. 1 Subchannel of a typical PWR. **a** Schematic view, **b** meshing surfaces generated in ANSYS fluent

Table 3 Design parameters of the subchannel of a typical PWR [12]

Parameters	Values (mm)
Length, l	600
Fuel rod diameter, d_f	9.5
Fuel rod pitch, p_f	12.54
Hydraulic diameter, d_h	11.8

The local and average heat transfer coefficient, average Nusselt number, and pressure drop associated with flow friction can be defined by the following formulae [4].

$$h(z) = \frac{q''}{T(z)_w - T(z)_b} \quad (8)$$

$$h_{av} = \frac{1}{l}\int_0^l h(z)dz \quad (9)$$

$$Nu = \frac{h_{av} d_h}{\lambda} \quad (10)$$

$$\Delta P = \frac{fl\rho v^2}{2d_h} \quad (11)$$

3.1 Boundary Conditions

A constant heat flux of 0.05 W/mm^2 is defined for the hydrodynamically smooth fuel rod surface. Fluid is assumed to enter the subchannel at 293 K. The inlet velocity (V_{in}) of nanofluid is varied along with the Reynolds number and nanoparticle concentration following the correlation $Re = (\rho v d_h/\mu)$. The range of Re is maintained within 20,000 and 80,000, while φ is altered from 1 to 4%. A relative average zero pressure is specified at the outlet of the subchannel. The length of the subchannel is taken to be long enough to obtain fully developed temperature and velocity profiles. Symmetry boundary condition has been applied for the fluid region.

3.2 Numerical Procedure

The computational fluid dynamics (CFD) model has been solved in ANSYS FLUENT software using the SIMPLER algorithm. Governing equations were discretized involving the second-order upwind scheme. Convergence criteria of the solution are set at residuals $\leq 10^{-7}$. A mesh sensitivity check has been performed for the single subchannel at Re = 80,000, φ = 4%, and the outcomes are presented in Table 4. A minor change has been observed in average Nusselt numbers for the mesh elements between 639,496 and 694,245. Hence, mesh elements of 639,496 have been chosen to conduct the numerical simulation. Moreover, the hexahedral mesh type is selected to attain better solution accuracy.

Table 4 Mesh sensitivity check on subchannel geometry at $Re = 80,000$ and $\varphi = 4\%$

Mesh elements	348,816	465,088	581,360	639,496	694,245
Nu	535.68	551.284	558.31	561.107	562.087
|Error|	–	2.83%	1.258%	0.498%	0.175%

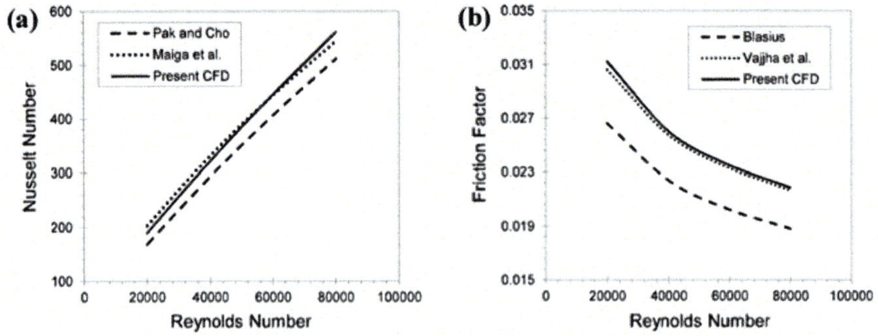

Fig. 2 Numerical code validation. **a** Nusselt number, **b** friction factor

3.3 Code Validation

For validating the CFD approach, a single subchannel holding Al_2O_3–TiO_2/water hybrid nanofluid has been considered. The average Nusselt number of the computational domain at $\varphi = 4\%$ has been compared with the correlations of Pak and Cho [13] experiment and Maïga et al. [14]. Moreover, the pressure drop/friction factor results data have been validated adopting the Blasius formula $\left(f = 0.3164Re^{-0.25}\right)$ and Vajjha et al. [15] experimental correlation. It is perceived from Fig. 2 that the results of the present numerical solution are congruent with the proposed correlations and predictions. However, comparatively higher deviations have been observed among the CFD data and the correlations derived from experiments [13, 15] as ideal conditions are considered in our study.

4 CFD Results

4.1 Temperature Profile

Temperature contours of the subchannel outlet for both pure water and hybrid nanofluid ($\varphi = 4\%$) at $Re = 80,000$ are presented in Fig. 3. Comparatively lower and improved temperature distribution has been observed for the hybrid nanofluid than its base fluid for a particular Reynolds number. The temperature profile across the subchannel midline (z-axis) under variable φ, and Re is illustrated in Fig. 4. Coolant is assumed to enter the subchannel at a predefined temperature, and the temperature rises steadily as it flows through the subchannel by absorbing heat from the fuel rod surfaces. Therefore, maximum coolant temperature is noticed at the subchannel outlet in all cases considered. The bulk temperature decreases with the increment in nanoparticle concentration and/or Reynolds number given that nanoparticle inclusion contributes to enhanced thermal conductivity, while increasing Reynolds number leads to higher inlet velocity.

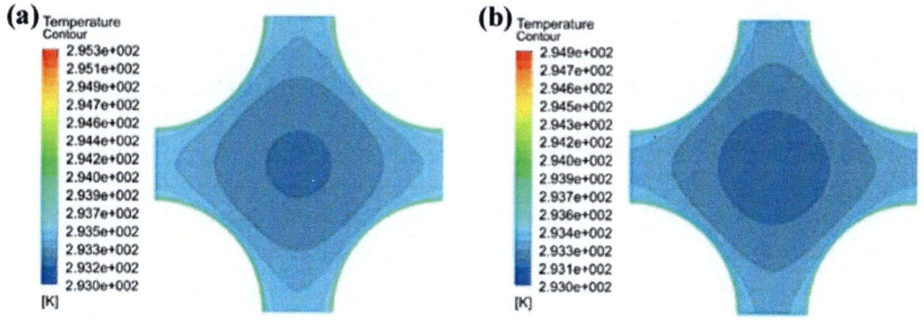

Fig. 3 Temperature contours of the subchannel outlet at $Re = 80,000$. **a** $\varphi = 0\%$, **b** $\varphi = 4\%$

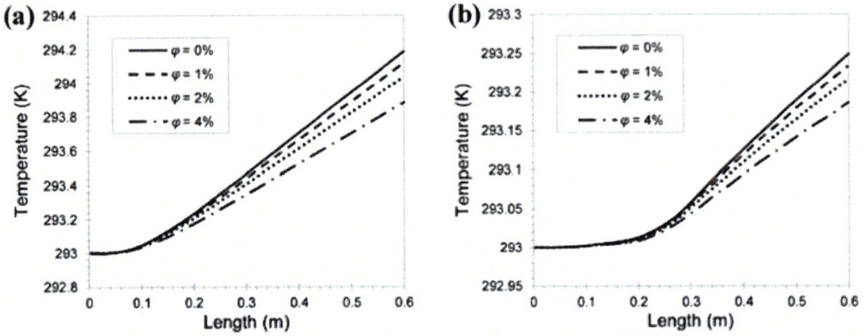

Fig. 4 Temperature profile across the subchannel centerline. **a** $Re = 20,000$, **b** $Re = 80,000$

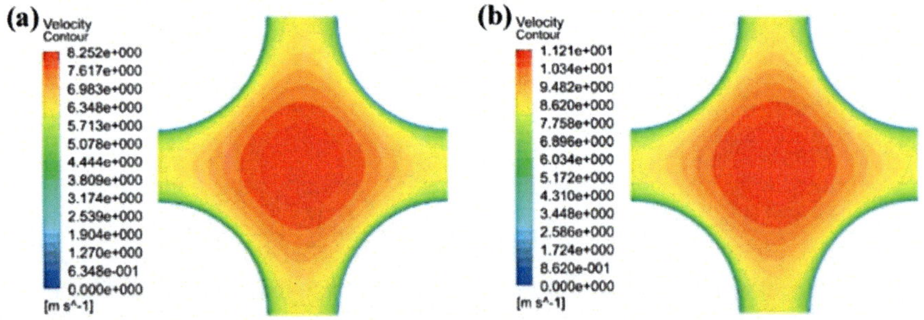

Fig. 5 Velocity contours of the subchannel outlet at $Re = 80,000$. **a** $\varphi = 0\%$, **b** $\varphi = 4\%$

4.2 Velocity Profile

Velocity contours of the subchannel outlet for 0 and 4% hybrid nanoparticle concentration at $Re = 80,000$ have been depicted in Fig. 5. At a particular *Re*, V_{in} varies along with the effective density and dynamic viscosity upon altering the nanoparticle volume concentration in a certain nanofluid. Hence, higher input and output velocities are noted with increasing φ than the pure liquid. However, the similar contour patterns at a fixed *Re* depict the proportionate change in the velocity profiles with respect to their corresponding inlet velocities at variable nanoparticle concentration.

From Fig. 6a, it has been perceived that the coolant centerline velocity rises continuously until the flow becomes hydrodynamically fully developed. For the computational domain under consideration, the fully developed regions have been marked at $z = 24d_{\text{h}}$ and $z = 28d_{\text{h}}$ approximately for $Re = 20,000$ and $Re = 80,000$, respectively. Moreover, the values of velocity ratio (V/V_{in}) reduce with the increment of Reynolds number on account of comparatively uniform

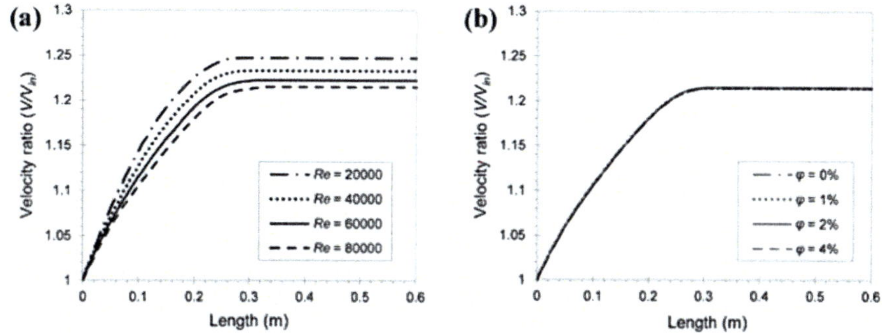

Fig. 6 Velocity ratio along the subchannel centerline (z-axis). **a** $\varphi = 4\%$, **b** $Re = 80,000$

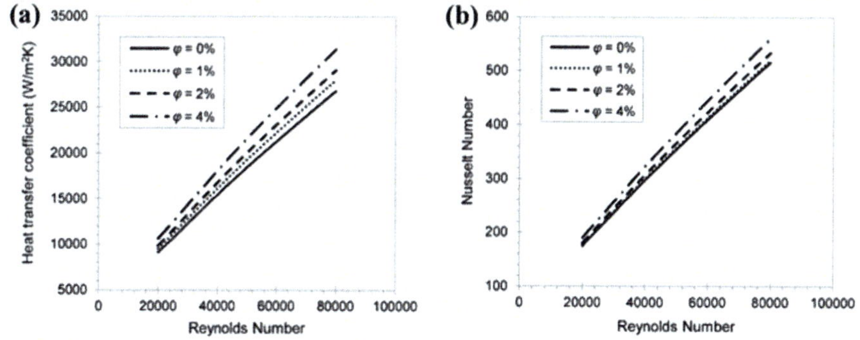

Fig. 7 **a** h versus Re graph, **b** Nu versus Re graph

velocity profiles at higher Re [4]. Furthermore, Fig. 6b indicates that the velocity ratio along the subchannel centerline is independent of φ at a specific Re under the assumptions considered in the conducted CFD study.

4.3 Heat Transfer

Figure 7 depicts a linear rise in h and Nu upon augmenting φ and Re within the computational domain under consideration. Though the outcomes of nanoparticle dispersion are inconsiderable at $\varphi = 1\%$, it becomes prominent at $\varphi = 4\%$. The CFD results confirm that the highest heat transfer rate takes place at the maximum Re for an individual nanoparticle content φ. Enhancement of heat transfer coefficient by 0.9%, 3.5%, and 17.2% and Nusselt number by 4.5%, 8.6%, and 8.8% has been identified for 1%, 2%, and 4% volume fraction of hybrid nanoparticles, respectively, at $Re = 80,000$ compared to pure fluid. This heat transfer exaltation in hybrid nanofluids can be ascribed to their superior thermophysical properties particularly the enhanced thermal conductivity upon Al_2O_3–TiO_2 hybrid nanoparticles inclusion [4, 9].

4.4 Pressure Drop

A linear decrease in coolant centerline pressure has been recorded along the length of the subchannel until it reaches the outlet with nearly zero pressure for all considered nanoparticle concentrations as illustrated in Fig. 8a. Figure 8b suggests an increase in pressure drop along with φ and Re inside the computational domain. Additionally, maximum pressure drops have been noticed at the uppermost *Re* for a particular φ. The coolant velocity elevation along the subchannel and enhanced viscosity at higher φ may be the leading contributors in suppressing the fluid pressure along its flow direction [11, 16]. The CFD results reveal pressure drop escalation by 17%, 42%, and 107% approximately for 1%, 2%, and 4% volume hybrid nanofluid, respectively, at $Re = 80,000$ with respect to the base fluid.

4.5 Thermal Performance Factor

Suspension of Al_2O_3–TiO_2 hybrid nanoparticle in pure water has emanated enhanced heat transfer coefficient in the subchannel coolant. Meanwhile, augmented viscosity has generated significant pressure drop which is undesirable and associated with energy loss. Hence, the thermal performance factor of the hybrid nanofluid should be evaluated to assess its compatibility for cooling operation. The thermal performance factor can be defined as follows [16]

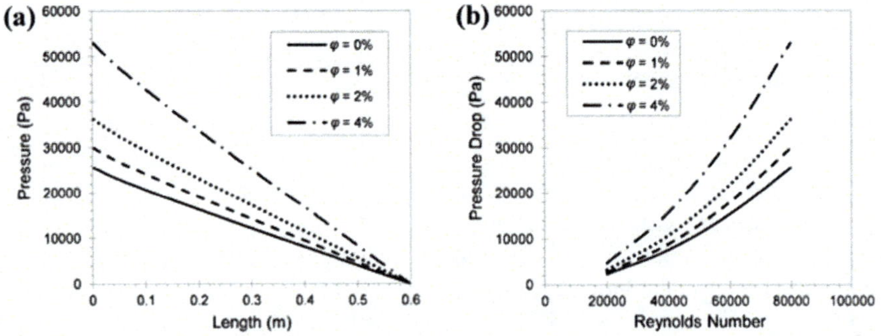

Fig. 8 **a** Coolant pressure distribution across the subchannel centerline at $Re = 80,000$, **b** relation between pressure drop and Reynolds number

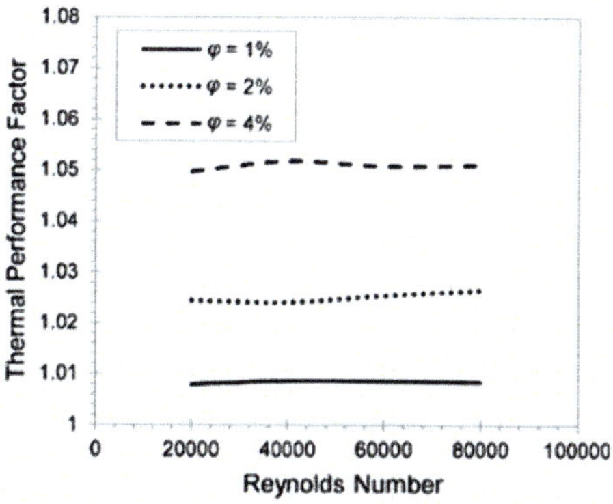

Fig. 9 Thermal performance factor versus Reynolds number graph

$$\text{Thermal performance factor} = \frac{(Nu_{\text{nf}}/Nu_{\text{f}})}{\sqrt[3]{(f_{nf}/f_{\text{nf}})}} \quad (12)$$

where thermal performance factor > 1.0 signifies energy gain.

The thermal performance factor of hybrid nanofluid against variable nanoparticle concentration and Reynolds number are found to be greater than one in all considered cases as presented in Fig. 9. Hybrid nanofluid at $\varphi = 4\%$ is perceived to be most economical with a performance factor = 1.05 approximately. Moreover, Fig. 9 suggests that *Re* impacts little the thermal performance factor for the cases considered. As pressure drop incurs a little penalty in pumping power, heat transfer amelioration within a PWR subchannel introducing Al_2O_3–TiO_2/water hybrid nanofluid seems to be promising. However, the neutron absorptivity of the proposed nanofluid must be investigated further to assess its feasibility in a nuclear reactor.

5 Conclusions

A CFD simulation is executed to predict the thermal transport of Al_2O_3–TiO_2/water hybrid nanofluid contained in a PWR subchannel under turbulent-forced convection conditions. The reasonability of hybrid nanoparticle inclusion in reactor coolant has been assessed with regard to the key parameters (i.e., heat transfer coefficient, pressure drop distribution, and thermal performance factor) recommended in IAEA Safety Reports Series No. 30. The conclusions drawn from the present study are as follows

1. Coolant centerline temperature is suppressed with the addition of hybrid nanoparticles in water attributed to thermal conductivity enhancement in a hybrid nanofluid. Maximum coolant temperature is always recorded at the outlet of subchannel geometry.

2. The flow becomes hydrodynamically fully developed at around $z = 24d_h$ and $z = 28d_h$ for $Re = 20,000$ and $Re = 80,000$, respectively. Nanoparticle volume fraction has no effect on coolant centerline velocity ratio (V/V_{in}) at a particular Reynolds number under the assumptions considered in the present CFD study.
3. Suspension of 1, 2, and 4% hybrid nanoparticle in pure water has contributed to the enhancement of heat transfer coefficient by 0.9, 3.5, and 17.2%, respectively, along with Nusselt number by 4.5, 8.6, and 8.8%, respectively, at $Re = 80,000$.
4. Linear coolant pressure drop is observed across the length of the subchannel. 1, 2, and 4% hybrid nanoparticle inclusion in the base fluid has resulted in pressure drop escalation by 17, 42, and 107%, respectively, at $Re = 80,000$.
5. Thermal performance factor > 1.0 in all instances under consideration imply that the application of Al_2O_3–TiO_2/water hybrid nanofluid is economical regardless of pressure drop penalty. However, neutron absorptivity investigation in the proposed nanofluid is recommended to assess its compatibility in a typical PWR.

References

1. Sajid, M.U., Ali, H.M.: Recent advances in application of nanofluids in heat transfer devices: a critical review. Renew. Sust. Energ. Rev. **103**, 556–592 (2019). https://doi.org/10.1016/j.rser.2018.12.057
2. Aladdin, N.A.L., Bachok, N.: Boundary layer flow and heat transfer of Al_2O_3–TiO_2/water hybrid nanofluid over a permeable moving plate. Symmetry **12**(7), 1064 (2020). https://doi.org/10.3390/sym12071064
3. Hadad, K., Kowsar, Z.: Twofold application of nanofluids as the primary coolant and reactivity controller in a PWR reactor: case study VVER-1000 in normal operation. Ann. Nucl. Energy **97**, 179–182 (2016). https://doi.org/10.1016/j.anucene.2016.07.008
4. Nazififard, M., Nematollahi, M., Jafarpur, K., Suh, K.Y.: Numerical simulation of water-based alumina nanofluid in subchannel geometry. Sci. Technol. Nucl. Ins. **2012**, (2012). https://doi.org/10.1155/2012/928406
5. Ghazanfari, V., Talebi, M., Khorsandi, J., Abdolahi, R.: Thermal–hydraulic modeling of water/Al_2O_3 nanofluid as the coolant in annular fuels for a typical VVER-1000 core. Prog. Nucl. Energ. **87**, 67–73 (2016). https://doi.org/10.1016/j.pnucene.2015.11.008
6. Bahrevar, M.H., Jahanfarnia, G., Pazirandeh, A., Shayesteh, M.: Thermal-hydraulic analysis of a novel design super critical water reactor with Al_2O_3 nanofluid as a coolant. J. Supercrit. Fluid **140**, 41–52 (2018). https://doi.org/10.1016/j.supflu.2018.05.029
7. Saadati, H., Hadad, K., Rabiee, A.: Safety margin and fuel cycle period enhancements of VVER-1000 nuclear reactor using water/silver nanofluid. Nucl. Eng. Technol. **50**(5), 639–647 (2018). https://doi.org/10.1016/j.net.2018.01.015
8. International Atomic Energy Agency (2003) Accident analysis for nuclear power plants with pressurized water reactors. Safety Reports Series No. 30, IAEA, Vienna
9. Kaska, S.A., Khalefa, R.A., Hussein, A.M.: Hybrid nanofluid to enhance heat transfer under turbulent flow in a flat tube. Case Stud. Therm. Eng. **13**, (2019). https://doi.org/10.1016/j.csite.2019.100398
10. Sharma, K.V., Sarma, P.K., Azmi, W.H., Mamat, R., Kadirgama, K.: Correlations to predict friction and forced convection heat transfer coefficients of water based nanofluids for turbulent flow in a tube. IJMNTFTP **3**(4), 1–25 (2012)

11. Oztop, H.F., Abu-Nada, E., Varol, Y., Al-Salem, K.: Computational analysis of non-isothermal temperature distribution on natural convection in nanofluid filled enclosures. Superlattice Microst. **49**(4), 453–467. https://doi.org/10.1016/j.spmi.2011.01.002
12. Liu, C.C., Ferng, Y.M.: Numerically simulating the thermal–hydraulic characteristics within the fuel rod bundle using CFD methodology. Nucl. Eng. Des. **240**(10), 3078–3086 (2010). https://doi.org/10.1016/j.nucengdes.2010.05.021
13. Pak, B.C., Cho, Y.I.: Hydrodynamic and heat transfer study of dispersed fluids with submicron metallic oxide particles. Exp. Heat Transfer **11**(2), 151–170 (1998). https://doi.org/10.1080/08916159808946559
14. Maïga, S.E.B., Nguyen, C.T., Galanis, N., Roy, G., Maré, T., Coqueux, M.: Heat transfer enhancement in turbulent tube flow using Al_2O_3 nanoparticle suspension. Int. J. Numer. Method H **16**(3), 275–292 (2006). https://doi.org/10.1108/09615530610649717
15. Vajjha, R.S., Das, D.K., Kulkarni, D.P.: Development of new correlations for convective heat transfer and friction factor in turbulent regime for nanofluids. Int. J. Heat Mass Tran. **53**(21–22), 4607–4618 (2010). https://doi.org/10.1016/j.ijheatmasstransfer.2010.06.032
16. Alshehri, F., Goraniya, J., Combrinck, M.L.: Numerical investigation of heat transfer enhancement of a water/ethylene glycol mixture with Al_2O_3–TiO_2 nanoparticles. Appl. Math. Comput. **369**, (2020). https://doi.org/10.1016/j.amc.2019.124836

Performance Analysis and Optimization of Ammonia-CO_2 and Ammonia–Propylene Refrigerant Pairs for Cascade Refrigeration

Harvendra Singh, Kaushalendra Kumar Singh, and Harshit Bahri

1 Introduction

Due to hazardous effects of halogenated refrigerants, Montreal Protocol (1987) initiated the phase out of CFC refrigerants. It was an agreement which was made in the year 1987 with the member countries of United Nations Industrial Development Organisations (UNIDO). Under this agreement, it was decided to phase down the use of chlorofluorocarbon by the end of nineteenth century. Before 1987, chlorofluorocarbon (CFCs), methyl chloroform, CTC Halon, and methyl bromide have been used as primary refrigerant. Kigali Amendment followed by many other amendments are done in this regard. As per agreement, the service of CFC refrigerants was terminated. Therefore, it is imperative to search for some alternative refrigerant which are environment-friendly. Many studies are done in search of alternative refrigerant couples. Due to environmental concerns caused by the use of HCFC and CFC refrigerants, natural refrigerants are again being established as the most important alternative refrigerants because of their inherent eco-friendly characteristics Parekh et al. [1]. Alberto et al. [2] have been designed the condenser for the cascade refrigeration system to operate eco-friendly refrigerant pair R (404) A-R (508) A; this refrigerant pair has zero ozone utilization likely and minimum global warming effect. The main objective of this model is to find out heat transfer coefficient of both condensers. Cascade system is generally designed to achieve temperature range up to −200 °C for the examination like cold storage, medical science in blood banks, and other freezing system Rezayan and Bahbaninia [3]. Sachdeva et al. [4] have been designed a model cascade refrigeration system on the basis of vapor compression cycle. In this study, we use working fluid in low temperature cycle in carbon dioxide (R744), ammonia (R717), propane(R290), propylene(R1270), R(404)A, and high-temperature cycle in which we use the R12

H. Singh · K. K. Singh (✉) · H. Bahri
Department of Mechanical Engineering, G L Bajaj Institute of Technology and Management, Greater Noida, Uttar Pradesh, India

L. M. Das et al. (eds.), *Recent Trends in Thermal Engineering*,
Lecture Notes in Mechanical Engineering,
https://doi.org/10.1007/978-981-16-3428-4_2

refrigerant. After that we are comparing the performance of the curve of NH_3, C_3H_8, C_3H_6, and R404A with R12, and we find out that ammonia is the best in comparing with the R12 Parke et al. [5]. Nasruddin and Arnas [6] used a mixture of CO_2 and hydrocarbon in LTC and recommended to use the mixture in a proportion of 63% ethylene 37% carbon dioxide in order to get the maximum effectiveness of the mixture refrigerant. Natural refrigerant blends are good alternatives to the solution of their low effectiveness Jadhav and Apte [7]. Low-temperature refrigeration systems obtained due to high pressure ratio and single-stage compression systems are not recommended because they require higher compressor power demand resulting in higher running cost Aminyavari et al. [8]. Lee et al. [9] performed a two-dimensional multi-objective optimization to find an optimum cascade condensing temperature which gives maximum COP and minimum exergy destruction. Bhattacharya et al. [10] worked on two-stage internally reversible cascade cycles to optimize the cooling load with respect to the intermediate temperature. Getu and Bansal [11] found the values of optimal evaporator temperature for a CO_2–NH_3 based cascade system in order to maximize its COP.

Instead number of studies regarding optimization and analysis of various refrigerant pairs in cascade refrigeration system, a comparative analysis based purely on natural refrigerant pairs has not been done so far in the reviewed literature. The aim of the study is to investigate the thermodynamic performances of cascade refrigeration system using NH_3–CO_2 and NH_3-propylene refrigerant pairs.

2 System Description

Figure 1 shows the representation of a cascade refrigeration system which is a series combination of two single-stage refrigeration systems, one is lower temperature cycle (LTC), and other is higher temperature cycle (HTC). Air from the cold space is fed to the evaporator where the cooling effect is obtained by LTC refrigerant evaporation. Then, the refrigerant is compressed by LTC compressor, and after condensation in cascade heat exchanger, it is fed back to the evaporator after undergoing expansion valve. The HTC refrigerant receives the heat in cascade heat exchanger, and similar processes are performed in HTC cycle. The heat from HTC refrigerant is rejected to the ambient environment in condenser through fan. Figure 2 shows the thermodynamic cycles of LTC and HTC on *T*-*s* diagram.

3 Mathematical Modeling

The cascade refrigeration system is modeled on EES software through thermodynamics equations based on energy and exergy balance equation. The program is validated with a previously published work. Energy analysis, exergy analysis, and

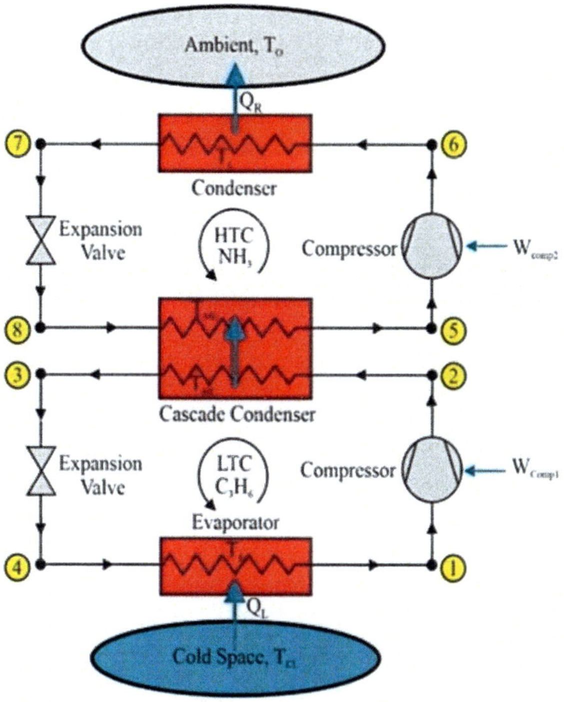

Fig. 1 Block diagram of cascade refrigeration system

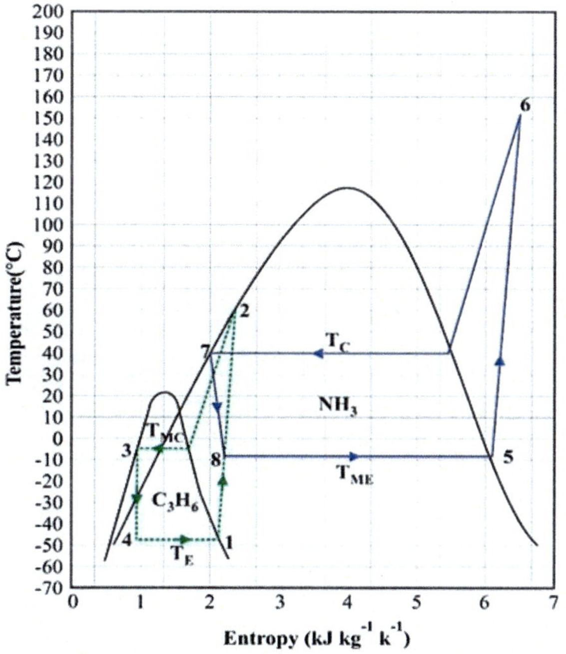

Fig. 2 Thermodynamic cycle on *T-s* diagram

optimization are performed to evaluate the optimal COP using various natural refrigerant couples. A comparative study is done to figure out the best refrigerant couple which reflects the best performance, i.e., maximum COP.

3.1 Energy Analysis

In energy analysis, the energy interactions between different system components are using energy balance equations. Following assumptions are involved in the analysis:

- No changes in heat losses and pressure.
- All components obey steady flow energy equations.
- No changes in potential and kinetic energy.
- Refrigerants are at respective saturated states at inlet and exit of heat exchangers.

Cooling load is:

$$\dot{Q}_{\text{L}} = \dot{m}_1(h_1 - h_4) \tag{1}$$

The isentropic efficiency (η_{isen}) of compressors can be written as [1]:

$$\eta_{\text{isen}} = 1 - (0.04 \times r_p) \tag{2}$$

The electrical efficiencies and mechanical (η_{e} and η_{m}) of both compressors (LTC and HTC) are assumed as 1.

LTC and HTC Compressor power inputs are expressed as:

$$\dot{W}_{\text{Comp}_1} = \frac{\dot{m}_1(h_{2s} - h_1)}{\eta_{\text{isen}}\eta_{\text{m}}\eta_{\text{e}}} = \frac{\dot{m}_1(h_2 - h_1)}{\eta_{\text{m}}\eta_{\text{e}}} \tag{3}$$

$$\dot{W}_{\text{Comp}_2} = \frac{\dot{m}_2(h_{6\text{s}} - h_5)}{\eta_{\text{isen}}\eta_{\text{m}}\eta_{\text{e}}} = \frac{\dot{m}_2(h_6 - h_5)}{\eta_{\text{m}}\eta_{\text{e}}} \tag{4}$$

Applying energy balances in expansion valves, we get:

$$h_3 = h_4 \tag{5}$$

and

$$h_7 = h_8 \tag{6}$$

Heat transferred in cascade heat exchanger expressed as:

$$\dot{Q}_{cc} = \dot{m}_2(h_5 - h_8) = \dot{m}_1(h_2 - h_3) \quad (7)$$

The heat rejected in condenser can be expressed as:

$$\dot{Q}_R = \dot{m}_2(h_6 - h_7) \quad (8)$$

Applying energy balance on overall system, we get:

$$\dot{Q}_H = \dot{Q}_L + \dot{W}_{Comp_1} + \dot{W}_{Comp_2} \quad (9)$$

Total work done on both compressors can be expressed by

$$\dot{W}_{Total} = \dot{W}_{Comp_1} + \dot{W}_{Comp_2} \quad (10)$$

Coefficient of performance (COP) expressed as

$$COP = \frac{\dot{Q}_L}{\dot{W}_{Total}} \quad (11)$$

3.2 *Exergetic Analysis.*

Exergy is the maximum obtainable work from the system when it is taken back to the dead state. Exergy destructed can be obtained by exergy balance equations applied to different components of the system:

$$\dot{E}_{d,evp} = \dot{Q}_L \cdot \left[1 - \frac{T_O}{T_{CL} + 273}\right] + \dot{m}_1.\dot{E}_4 - \dot{m}_1.\dot{E}_1 \quad (12)$$

$$\dot{E}_{d,comp,1} = \dot{m}_1.\dot{E}_1 + \dot{W}_1 - \dot{m}_1.\dot{E}_2 \quad (13)$$

$$\dot{E}_{d,cc} = \dot{m}_1.\dot{E}_2 + \dot{m}_2.\dot{E}_8 - \dot{m}_2.\dot{E}_5 - \dot{m}_1.\dot{E}_3 \quad (14)$$

$$\dot{E}_{d,exp,1} = \dot{m}_1.\dot{E}_3 - \dot{m}_1.\dot{E}_4 \quad (15)$$

$$\dot{E}_{d,comp,2} = \dot{m}_2.\dot{E}_5 + \dot{W}_2 - \dot{m}_2.\dot{E}_6 \quad (16)$$

$$\dot{E}_{d,cond} = \dot{m}_2.\dot{E}_6 - \dot{m}_2.\dot{E}_7 \quad (17)$$

$$\dot{E}_{\text{d,exp,2}} = \dot{m}_2.\dot{E}_7 - \dot{m}_2.\dot{E}_8 \quad (18)$$

The total inlet exergy into the system can calculated as:

$$\dot{E}_{\text{in}} = \dot{W}_{\text{Comp1}} + \dot{W}_{\text{Comp2}} \quad (19)$$

The total exergy out of the system expressed as:

$$\dot{E}_{\text{out}} = \dot{Q}_{\text{L}} \cdot \left[\frac{T_{\text{O}}}{T_{\text{CL}} + 273} - 1 \right] \quad (20)$$

Total destruction in exergy can be written as:

$$\dot{E}_{\text{d,Total}} = \dot{E}_{\text{in}} - \dot{E}_{\text{out}} = \sum_{\text{p}} \dot{E}_{\text{d,p}} \quad (21)$$

where $\sum_{\text{p}} \dot{E}_{\text{d,p}}$ is the summation of all exergy destructions.

Exergetic efficiency can be expressed as:

$$\eta_{\text{II}} = \frac{\dot{E}_{\text{out}}}{\dot{E}_{\text{in}}} = 1 - \frac{\dot{E}_{\text{d,Total}}}{\dot{E}_{\text{in}}} \quad (22)$$

3.3 System Optimization

Optimization is the way toward finding the best or least estimation of a capacity for some imperative, which must be genuine paying little heed to the arrangement. Alternatively, it means the best possible solution for a given problem under defined set of constraints.

In this study, we have done optimization using engineering equation solver (EES) software. In this study, the objective function is COP. EES software consist of optimization tool which is used for optimization of cascade refrigeration system (using C_3H_6–NH_3 refrigerant couple). The method has used for optimization is "conjugate direct method." The following decision variable is presented in Table 1. The list of the input parameter used for cascade refrigeration system is presented in Table 2.

Table 1 Range of important parameters assumed in simulation work [8]

Design variables	Range
LTC evaporator temperature	−56 °C < T_E < −47 °C
LTC condensation temperature	−11 °C < T_{MC} < 1 °C
Cascade temperature difference	2 °C < ΔT_{CC} < 10 °C
HTC condensation temperature	40 °C < T_{MC} < 65 °C

Table 2 Other important parameters assumed in system modeling [8]

Parameter	Value
Load of refrigeration ($\dot{Q}_L$)	50 KW
Environment temperature (T_o)	25 °C
Ambient pressure (P_o)	1 atm
Cold space temperature (T_{CL})	−45 °C

4 Results and Discussions

4.1 Model Validation

Simulation model of the cascade refrigeration system has been validated through a previously published work [8]. Published values are compared with the values calculated from the present model, and the respective deviations in the values are given in Table 3.

As appeared in Table 3, the accomplished qualities are in acceptable concurrence with the distributed information in the wake of ascertaining the deviation *D* from following equation:

$$D = \frac{R - M}{R} \times 100 \tag{23}$$

4.2 Effect of Design Parameters on Exergetic Efficiency and COP

In this study, we have calculated the COP and exergetic efficiency for cascade refrigeration system using both refrigerant couple (CO_2–NH_3 and C_3H_6–NH_3). The design variable alongside their relating scope of varieties (as appeared in Table 1) are utilized for plotting graph as shown below.

Variation of COP and Exergetic efficiency with T_E

Figure 3 appeared the variation in COP and exergetic efficiency with variation in evaporator temperature (T_1). By varying the temperature of evaporator from −47 to −56 °C, from the graph, we can see that as the temperature increases, the COPs and exergetic efficiencies also increase both refrigerant, but the COP and exergetic

Table 3 Validation of present model from Ref. [8]

Operating parameters				COP			Exegetic efficiency		
T_E (°C)	T_{MC}(°C)	ΔT_{CC} (°C)	T_C (°C)	*R*	*M*	*D*	*R*	*M*	*D*
−47	−6.301	40	2	1.4949	1.56	4.35	45.89	47.89	4.35

R reference, *M* model, *D* deviation (%)

efficiency of propylene–ammonia refrigerant is more as compared to carbon dioxide-ammonia refrigerant throughout the graph for the same temperature difference. As a result, we can say that propylene–ammonia refrigerant is more efficient than carbon dioxide–ammonia refrigerant.

Variation of COP and exergetic efficiency with LTC condenser Temperature

Figure 4 is similar to Fig. 3. The only difference is that it shows the variation in exergetic efficiency and COP with variation in LTC condenser temperature (T_3). By varying the temperature of LTC condenser from −11 to 1 °C, we can observe from the graph that the effect of LTC condenser temperature on COPs and exergetic efficiencies is similar in both refrigerants, the only difference is that the exergetic efficiency and COP of propylene–ammonia refrigerant is more as compared to carbon dioxide–ammonia refrigerant throughout the graph for the same temperature difference. As a result, we can say that propylene–ammonia refrigerant is more efficient than carbon dioxide–ammonia refrigerant.

Variation of COP and Exergetic efficiency with HTC condenser Temperature (T_7)

Figure 5 shows the variation in exergetic efficiency and COP with variation in HTC condenser temperature (T_7). This graph shows the influence of HTC condenser temperature (T_7) on COP and exergy efficiency. By varying the temperature of HTC condenser from 40 to 65 °C, it is clear from the graph that, with the increase in

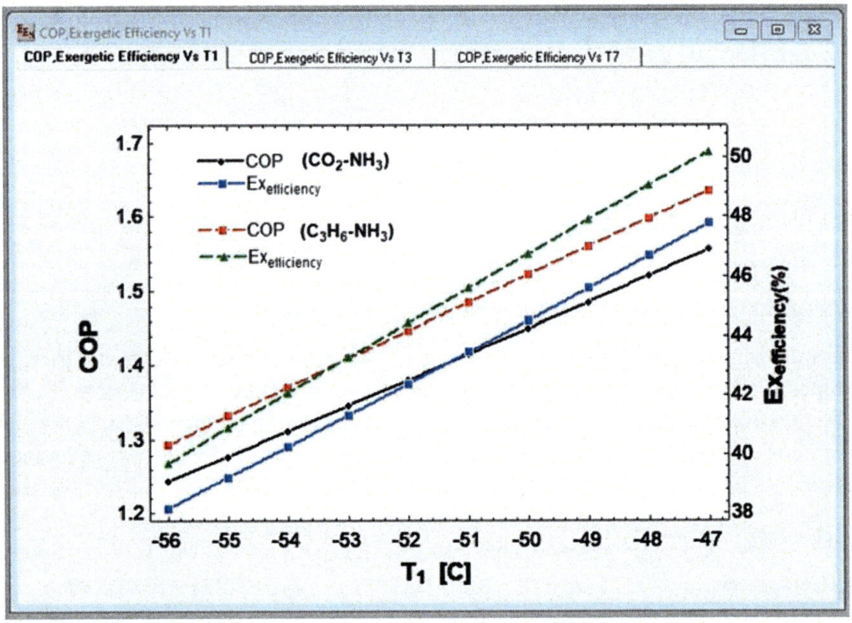

Fig. 3 Variation of COP and exergetic efficiency with T_E of both refrigerant couple (CO_2–NH_3 and C_3H_6–NH_3)

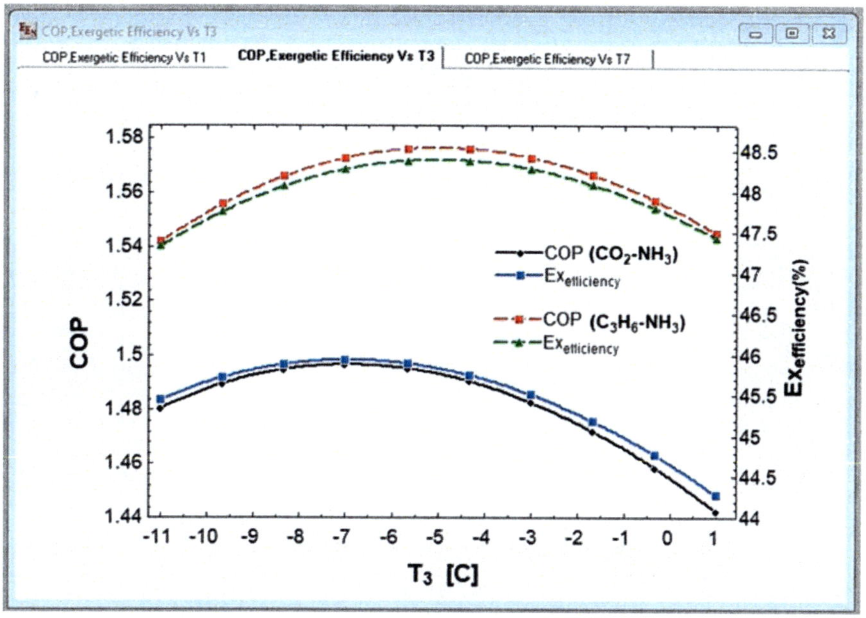

Fig. 4 Variation of COP and exergetic efficiency with LTC condenser temperature for both refrigerant couple (CO_2–NH_3 and C_3H_6–NH_3)

temperature, COP and exergy efficiencies decrease for both refrigerant, but the exergetic efficiency and COP of propylene–ammonia refrigerant are more as compared to carbon dioxide–ammonia refrigerant throughout the graph for the same temperature difference. Therefore, we can say that propylene–ammonia refrigerant is more efficient than carbon dioxide–ammonia refrigerant.

4.3 Optimization Results

A single objective optimization of the system using CO_2–NH_3 refrigerant pair to maximize the exergetic efficiency has been done using conjugate direct method. The optimization results of CO_2–NH_3 refrigerant couple are given in Table 4.

Table 4 presents the results of optimization for CO_2–NH_3 refrigerant pair. It is evident that the evaporator temperature is −47 °C which is almost the minimum value of its provided range. This is because the system performance is maximum when the temperature gap between condenser and evaporator is minimum which results in minimization of the power required. The maximum exergetic efficiency is 47.89% and minimum compressor work is 14.19 kW.

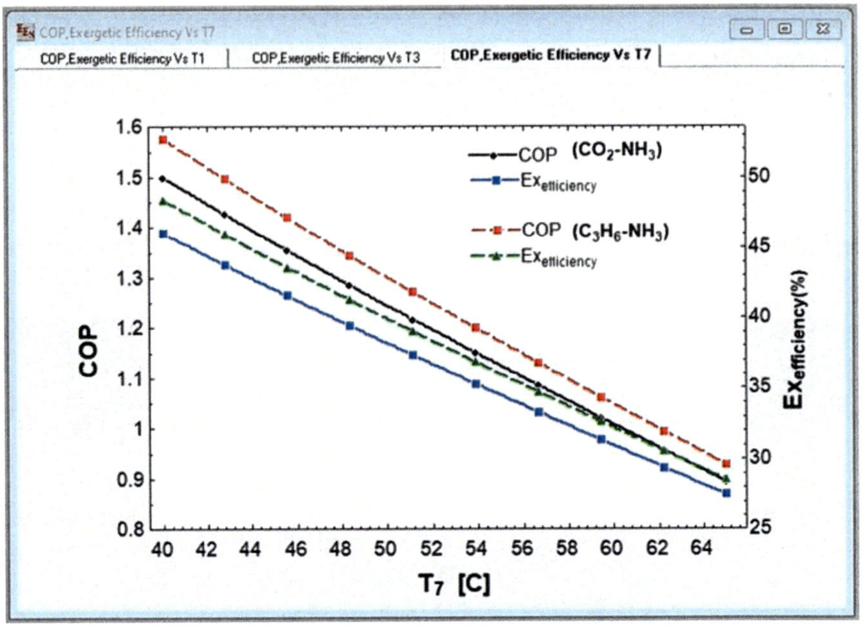

Fig. 5 Variation of exergetic efficiency and COP with HTC condenser temperature for both refrigerant couple (CO_2–NH_3 and C_3H_6–NH_3)

Table 4 Results of optimization of the carbon dioxide–ammonia refrigeration system

Optimal operating conditions				Optimum performance parameters			Total exergy destruction (KW)
T_E (°C)	T_{MC} (°C)	ΔT_{CC} (°C)	T_C (°C)	$(COP)_{max}$	Exergetic efficiency (%)	Compressor work (kW)	
−47	−6.301	2	40	1.56	47.89	14.19	16.7

5 Conclusion

After carrying out the comparative analysis and optimization of cascade refrigeration system using NH_3–CO_2 and NH_3-propylene refrigerant combinations, following conclusions are drawn from the study:

1. Values of exergy efficiency and COP increase with the rise in evaporator temperature for both the refrigerant pairs.
2. Values of exergy efficiency and COP first increase and then start to decrease with an increase in LTC condenser temperature.
3. Exergy efficiency and COP values continuously reduce with rise in HTC condenser temperature for both refrigerant couples.

4. C_3H_6–NH_3 refrigerant couple is more efficient than CO_2–NH_3 refrigerant couple.

References

1. Parekh, A.D., Tailor, P.R., Jivanramajiwala, H.R.: Optimization of R507A-R23 cascade refrigeration system using genetic algorithm. Int. J. Mech. Mechatron. Eng. **4**, 10 (2010)
2. Alberto, J., Dopazo, J., Fernández-Seara, Sieres, J., Uhía, F.J.: Theoretical analysis of a CO_2–NH3 cascade refrigeration system for cooling applications at lower temperatures. Appl. Therm. Eng. **29**, 1577–1583 (2009)
3. Rezayan, O., Behbahaninia, A.: Thermoeconomic optimization and exergy analysis of CO_2/NH3 cascade refrigeration systems. Energy **36**, 888–895 (2011)
4. Sachdeva, G., Jain, V., Kachhwaha, S.S.: Performance study of cascade refrigeration system using alternative refrigerants. Int. J. Mech. Aerosp. Ind. Mech. Manufact. Eng. **8**, 3 (2014)
5. Parke, H., Kim, D.H., Kim, H.S.: Thermodynamic analysis of cascade refrigeration system for optimal intermediate temperatures in R134aeR410A. Appl. Therm. Eng. **54**, 319–327 (2013)
6. Nasruddin, S.S., Arnas, N.G.: Optimization of a cascade refrigeration system using C_3H_8 in high temperature circuit (HTC) and a mixture of C_2H_6/CO_2 in low temperature circuit (LTC). Appl. Therm. Eng. **104**, 96–103 (2016)
7. Jadhav, J.S., Apte, A.D.: Review on cascade refrigeration system with different refrigerant pairs. International journal of innovations in engineering research and technology. IJIERT **2** (6) (2015). ISSN: 2394–3696
8. Aminyavari, M., Najafi, B., Shirazi, A., Rinaldi, F.: Exergetic, economic and environmental (3E) analyses, and multi-objective optimization of a CO_2/NH_3 cascade refrigeration system. Appl. Therm. Eng. **65**, 42–50 (2014)
9. Lee, T.S., Liu, C.H., Chen, T.W.: Thermodynamic analysis of optimal condensing temperature of cascade-condenser in CO_2/NH_3 cascade refrigeration systems. Int. J. Refrig. **29**, 1100–1108 (2006)
10. Bhattacharyya, S., Bose, S., Sarkar, J.: Exergy maximization of cascade refrigeration cycles and its numerical verification for transcritical CO_2–C_3H_8 system. Int. J. Refrig. **30**, 624–632 (2007)
11. Getu, H.M., Bansal, P.K.: Thermodynamic analysis of an R744–R717 cascade refrigeration system. Int. J. Refrig. **31**, 45–54 (2008)

A Comprehensive Review of Performance, Combustion, and Emission Characteristics of Biodiesel-Fueled Diesel Engines

Suraj Bhan, Raghvendra Gautam, Pusphendra Singh, and Abhishek Sharma

Nomenclature

ASTM	American standard test method
B10KOEE90	Diesel 90% + Kusum oil ethyl ester
B15KOEE95	Diesel 85% + Kusum oil ethyl ester 15%
B20KOEE80	Diesel 80% + Kusum oil ethyl ester 20%
B5KOEE95	Diesel 95% + Kusum oil ethyl ester 5%
BEMP	Brake effective mean pressure
BP	Brake power
BSFC	Brake specific fuel consumption
BTE	Brake thermal efficiency
CFPP	Cold filter plugging point
CI	Compression ignition
CN	Cetane number
CO	Carbon mono oxide
CO_2	Carbon dioxide
CP	Cloud point
CRDI	Common rail direct injection
CV	Calorific value
D	100% diesel
F/A ratio	Fuel-air ratio
FAME	Free acid methyl ester
FFA	Free fatty acids
FIP	Fuel injection pressure
JOEE5	Jatropha oil ethyl ester 10% + 90% diesel

S. Bhan (✉) · R. Gautam · P. Singh
Department of Mechanical Engineering, Delhi Technological University, Delhi, India

A. Sharma
Department of Mechanical Engineering, G L Bajaj Institute of Technology and Management, Greater Noida, Uttar Pradesh, India

L. M. Das et al. (eds.), *Recent Trends in Thermal Engineering*,
Lecture Notes in Mechanical Engineering,
https://doi.org/10.1007/978-981-16-3428-4_3

JOEE5	Jatropha oil ethyl ester 5% + 95% diesel
JOME10	Jatropha oil methyl ester 10% + 90% diesel
JOME5	Jatropha oil methyl ester 5% + 95% diesel
KOH	Potassium hydroxide
MAG	Mono-acylglycerols
NaOH	Sodium hydroxide
NOx	Oxides of nitrogen
PM	Particulate matters
TAG	Triacylglycerols
TDC	Before top dead center
UHC	Unburned hydrocarbons

1 Introduction

Expanding attention to the consumption of petroleum derivative fuel as well as their negative natural effects has set off intrigue in the likely advantages of biofuels, for example, biodiesel, which may be the promising elective fuel for diesel engines. Biodiesel is beneficial in many aspects like as it is biodegradable, sustainable, and compatible to use in diesel engines without modification. It has a high flash compared to petroleum fuel, so that it is easier to handle. It was seen that biodiesel produces less harmful emissions like being CO_2, CO, unburned hydrocarbons (UHC), and PM comparative with petro-diesel [1].

Biofuel molecular structure contains more oxygen in comparing fossil fuel due to which it contains more double bond comparatively. Biofuel contains saturated and unsaturated fatty acids. Unsaturated fatty acids are responsible to increase the cetane number and viscosity of the fuel. Unsaturated fatty acids contained double bonds which confined less space in comparison with single bonds resulting in density and viscosity increased in comparison with fossil fuel. Biofuel has a higher viscosity, density, and bulk modulus compared to fossil fuel [2].

Biodiesel is produced from vegetable oil, animal fats, and certain lipid-rich algae species. The principal molecular part of oils and fats are fatty substances, otherwise called triacylglycerols (TAG). There are many conventional methods for producing biofuels, such as pyrolisis, microemulsion, dilution, and thermal cracking. (Table 1)

But in recent technology, transesterification is proving to be a better way to produce biofuels. In transesterification reactions, free fatty acids of triglycerides are converted to alkyl ester. In the transesterification reaction, there are three main continuous reactions, of which diglyceride and monoglycerides are the main intermediate products. The transesterification process reduces overall viscosity without changing the calorific value of the fuel. During the transesterification process, free fatty acids from biofuels are converted into mono-alkyl esters and glycerol in the presence of alcohol such as methanol or ethanol. These reactions can be catalyzed by acid, alkali, and enzyme catalysts to increase the speed, efficiency,

Table 1 Various types of methods of biofuel production [3–5]

Production methods	Advantages	Shortcoming
Pyrolysis	Very simple, eco-friendly process (non-polluted)	Need high temperature, costly apparatus, less pure due to present of carbon residues
Micro-emulsion	Simple process	Produce high-viscous biofuel, and biofuel is also poor in stability and volatility
Blending	Simple process	Poor combustion produce carbon oxides and particles
Reactive distillation	Beneficial for the high free fatty acid biofuel. Less use of methanol and easily separation of by-products	High required amount of energy. Mass transfer depends on catalyst performance
Microwave technology	Less reaction time, less amount of heat loss	It is not suitable for feedstock with solids. Requirement of catalyst separation from end product. Conversion depends upon catalyst performance
Super critical method	Catalyst-free reaction, high speed of reaction conversion, highly conversion	High amount of energy consumes. Process takes place at the critical temperature
Transesterification	Highly conversion, rapid reaction rate, can be used with or without catalyst. It can be commercially used	Cost catalyst selection depends on type of feedstock

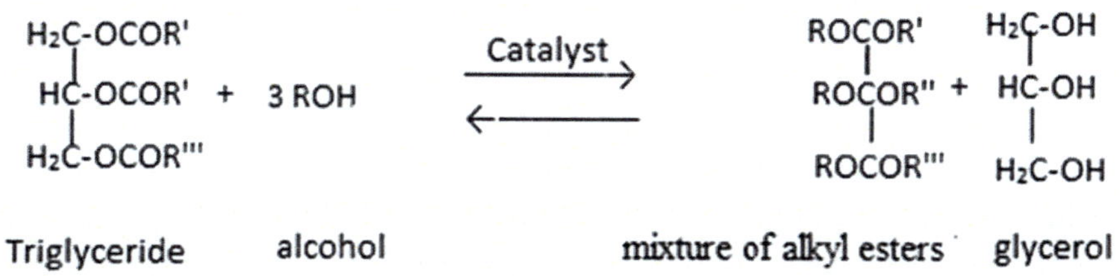

Fig. 1 Transesterification reaction of biofuel [3]

and yield of biofuels. These catalysts may be heterogeneous, homogeneous, and enzymes, depending on the type of raw material [3–5].

The transesterification process is given in Fig. 1.

The transesterification reaction is an arrangement of three persistent and reversible reactions. These reactions proceed stepwise. In the first starting step, di-acylglycerols and mono-acylglycerols (MG) are formed. These are known as intermediates products. In the last step, glycerol is produced [6, 7]. The transesterification response must be adequately finished and the FAME purged to satisfy fuel quality guidelines, for example, EN 14214. Biofuel feedstock have more viscosity than conventional biofuel. Transesterification is a common method that is used to reduce the viscosity of biofuel close to diesel fuel. The transesterification process converts the triacylglycerols into free. Without transesterification, higher viscosity biofuel makes the poorer atomization of the ignition cylinder worse and the ignition device improper coming about knocking and so on operational issues in diesel engine, for example, deposit formation on engine parts, increase specific fuel combustion, larger ignition delay, improper combustion resulting knocking, and possible engine failure [8]. TAGs high viscosity, density, high pour point, and low volatility can lead to extreme engine deposition, injector coking, and cylinder ring stickiness. To meet the fuel's standard viscosity, biofuels are mixed in diesel fuel [2]. Although, triacylglycerol used in a diesel engine without modification. But reduced viscosity biofuel utilization is efficient in engines because higher density biofuels are usually responsible for engine deposition, cylinder ring sticking and injector coking, etc. [9]. Biofuel has some advantage in some aspect like as:-

- Biofuel has physical and chemical properties that are similar to diesel.
- It may be locality available in various types of feedstock. It may be produced from waste fatty material. Although biofuel is produced, solid waste management may occur on other aspects.
- It is compatible to use in a diesel engine without modification in diesel engine design.
- It is miscible with diesel due to which it can use with a blend of diesel.
- Biofuel is biodegradable, sustainable, non-toxic, and eco-friendly.
- Biofuel is oxygenated fuel due to which it offers more oxygen during the combustion of fuel. Oxygen molecules help in the burning of fuel. So that proper burning of fuel takes place inside the combustion chamber.
- Biofuel flash point is high, so that it is easy to handle in the transport system.
- Biofuel is more viscous in comparison with diesel due to which fantastic lubricity to diminish wear and to expand the life of fuel injection.
- Biodiesel holds extraordinary potential for animating economical country improvement and an answer for energy security issues.
- Biodiesel should not be transported, refined, and drilled like diesel.
- Biodiesel creation is simpler in contrast with diesel.
- Biodiesel has a higher cetane number in comparison with diesel.
- It has the capacity to diminish CO_2 discharges contrasted with fossil diesel or remaining CO_2 nonpartisan [10].

On the other aspect, biofuel has some limitations like as:-

- Biofuel has a lower calorific value in comparison with diesel fuel.
- It has a high pour point, so that it has poor storage stability and cold flow properties.
- Biofuel has lower oxidation stability.
- Biodiesel has a corrosive nature against copper and brass.

- Biodiesel produces lower engine speed and torque on the shaft. The biodiesels on the normal abatement power by 5% contrasted with that of diesel at rated load.
- Coking of injectors on the piston and head of the engine is the major problem. Researchers are attentive toward it to remove this problem.
- Biofuel has high viscosity due to which it has inferior spray characteristics resulting in insufficient mixing of fuel, improper combustion of fuel.
- Biodiesel degradation under storage for prolonged periods.
- In the case of biofuel combustion of fuel take place at higher temperature due to more oxygen biofuel molecule in the structure of biofuel.
- The high viscosity, in long haul activity, presents the arrangement of injector deposits, stopping of channels, lines and injectors, ring sticking, and contradiction with ordinary.
- It was seen by the researcher that biofuel has the more BSFC and lower BTC. But it can say that biofuel produced lesser CO and HC emissions but increase CO_2 emissions comparatively. At the higher flame temperature, oxides of nitrogen formed.
- Biodiesel causes excessive engine wear.

These weaknesses can be overwhelmed by appropriately picking the feedstock of biofuel. Transesterification is appropriate method to produce the biofuel from various feed stock. Transesterification method is easy and cheap method which is used to reduce to viscosity of biofuel and ultimately produce biofuel. It is likewise announced that the utilization of biodiesel builds NO_X emissions [11]. There are many reasons for increase in NO_X discharge which are given below:-

- Higher flame temperature
- Higher density
- Fuel injection advancement
- Lower volatility
- Higher iodine value
- Presence of oxygen molecule in fuel structure

It is additionally seen that biodiesel powers are fit for diminishing soot emissions [12, 13] because of nonappearance of sulfur furthermore, aromatics [14] and presence of oxygen molecule in structure of biofuel. Consequently, it is basic to consider the study, combustion, and emissions attributes of biodiesel fills to embrace them as substitute fuel for diesel engines.

This literature review provides the effective information about the impact of biodiesel on engine performance like as brake power, fuel consumption, brake thermal efficiency, and emissions, including: engine emissions, such as nitrogen oxides (NOx), carbon monoxide (CO), particulate matter (PM), hydrocarbons (HC), carbon dioxide (CO_2), and smoke opacity. Engine performance and engine are affected from the various type of the feed stock because of biodiesel physical properties influenced by the saturated and unsaturated fatty acid of fuel. Different types of biofuel feedstock have the slightly change in amount of saturated and unsaturated fatty acids. So it is difficult to conclude the emissions of biofuel. Although it was usually seen by the researcher, there is an increase in oxides of nitrogen, decrease in CO, particulate matter, unburned hydrocarbon, and aromatics.

2 Effect of Biofuel Blend on Performance and Emission Characteristics of C.I. Engine

Many researchers have studied biofuel emissions. They found that when biofuels are mixed with petroleum fuel, emissions such as HC, carbon dead oxides, and CO, carbon dioxide is reduced to one level, but oxides of nitrogen and particulate matter usually increase. This is a major problem with biofuels. Nitrogen oxides are very dangerous particles that cause pollution. Therefore, it is necessary to find ways to reduce nitrogen oxide and flue gas emissions.

NOx emissions depend on the main factors, such as increased flame temperature, the initial thrust of injection pressure, higher density, low volatility, high iodine value, a higher volume of modulus, and higher octane value. Researchers are looking at various factors and studying them to find solutions to reduce nitrogen oxides.

Researchers study the emission characteristic of JOME and JOEE blending with diesel fuel. JOME and JOEE blending fuel contain more oxygen compared to pure diesel, so that blended fuel has a more cetane number. So combustion of fuel takes place at a higher temperature inside the cylinder. At higher temperatures more NOx produced. So that it was seen that at full-load conditions, oxides of nitrogen increase 34% for JOME20 and 32% for JOEE20 in comparison with pure diesel. For another blending, there was not much increase in NO_X emissions. Oxygen molecule increases due to mixing of biofuels like JOEE and JOME. Oxygen molecule helps in proper combustion and oxidation of fuel. So that it was seen by the researcher that CO, CO_2, and HC are lower for all blends at all load conditions in comparison with pure diesel [15].

Brake thermal efficiency is defined as the ratio of brake power energy to the input fuel energy. It was found that BTE is less for all blending of KOEE with butanol in comparison with pure base diesel line. KOEE biofuel has the lower calorific value and less volatility. But it was seen that for the various blending like as the B5KOEE95, B10KOEE90, B15KOEE85, and B20KOEE80, BTE are 26.67%, 27.45%, 29.67%, and 29.00%. When butanol is mixed with the KOEE, oxygen contents increased resulting proper combustion and air-fuel mixing takes place inside the combustion chamber. So BTE increased. But KOEE has less BTE due to higher viscosity which leads the improper atomization. It was seen that at all loads for all blend BSFC is higher than diesel base line. This was due to lower calorific value fuel required more consumption to produce the same output. It was seen that at 40–60% load for all blend, CO and HC emissions are lower in comparison with the diesel base line which may be due to higher oxygen content in the fuel. As load increased, CO emissions increased due to more amount of fuel injected inside the cylinder due to which improper combustion take place. But HC emissions are found lower for all blending and all loading conditions.CO_2 emissions found lightly higher than the neat kusum oil when butanol mixed with biofuel. It was seen that NO_x are lower for all KOEEB blends at high load which may be due to quenching effect of butanol and lower flame temperature [16].

When lower alcohol biofuel like as methanol and ethanol are blended with the diesel, emissions characteristics like as CO and HC improved. Alcohol biofuels are enriched in oxygen molecule due to which they give the better combustion characteristics. NO_X emission also improved which may be due to combustion of fuel take place at higher temperature. But it was found that higher alcohol gives the better results in comparision with the lower alcohol like as when n-octonal mixed with the diesel, it gives the better BTE, lower BSFC, and less NO_X emissions in comparision with lower alcohol blending. It was found that mostly biofuel blending near about to 20–30% gives optimum performance characteristics [17].

It was seen by the researcher when bio-ethanol is blended with the biodiesel kinematic viscosity, cetane number, lubricity, and corrosion resistance decreased. Higher viscosity problem can be easily eliminated. Due to higher viscosity, poor atomization of fuel takes place resulting improper combustion. Ethanol contains the 30% oxygen which improves the performance and combustion characteristics [18].

Researcher analyzes the test results and finds out that advanced injection timing is suitable for the alternative natural gas. It was seen that fuel consumption slightly increases with the advanced injection timing. There was a noteworthy decrease in CO_2 emissions with advanced injection timing. With the advanced injection timing, the CO emissions were significantly reduced in comparision with standard timing. The HC discharges of the double fuel frameworks were high all through the loading conditions. A further 1.51 development would in general deliver flighty conduct of the engine. It was seen that at the high load combustion chamber, temperature increases due to which evaporation rate of the fuel increased resulting reduce ignition delay. So that advanced injection is not suitable at high load conditions. Delay period directly affects the emissions and performance characteristics of the engine [19].

The main emission particulates are the oxides of nitrogen, carbon monoxide, unburned hydrocarbon emissions, and smoke. When fish oil is blended as biofuel with diesel and its blends B25, B50, B75, and B100 are used as fuel, the exhaust emissions like as smoke density increases in comparision with diesel. When biofuel mixes with diesel, its viscosity increases resulting poor atomization of fuel and also affects the air fuel mixing. This is due to improper combustion of the fuel inside combustion chamber. It was observed that when fish oil blended with diesel produce higher NO_X emissions in comparision with diesel. However, it is compared with the lower mixture of the fish oil biodiesel with diesel (B75) and B100 produces lower NO_X emissions. The higher NOX emission of fish oil biodiesel is due to its earlier combustion start compared with diesel. But emissions like as CO and HC increases with the blending of biofuel. Although it was seen that B100 fuel produced lower HC emissions than the diesel, higher HC and CO means that complete combustion of fuel does not take place inside the combustion chamber [20].

It was found that when the biofuel neem oil blended in the diesel, the BTE decreases to an extent. It may due to viscosity of the fuel increases in comparision with diesel due to which poor atomization of fuel takes place inside the combustion chamber raised resulting improper combustion. But it was seen that DN5 showed better peak load BTE than the D100. BSFC increases with increasing the blend of neem oil. But it was seen that minimum BTE is at 5% blend of neem oil [21].

It was seen by the researcher that methanol and ethanol are effective to use in transesterification process. But it was seen by the researcher that higher alcohol fuel gives the better result in comparision with the lower alcohols. So that n-octanol shows extraordinary potted-tial-and-diesel mixtures because of their high cetane number. The survey showed that the lower alcohol mixing indicated a decrease in BTE due to the high oxygen content of the fuel and the better combustion characteristics. In addition, BSFCs concurrent increment and lower braking power due to the low thermal value of biofuels. Alcohol-based fuel is higher oxygenated fuel. They give the additional oxygen and improve the combustion of fuel. Due to complete combustion of fuel CO and UHC decreased, it was seen that NOx emissions decreased when alcohol blended quantity increase with diesel fuel which is because of lower temperature of alcohol. It can moreover be done up for a greater part of the alcohol mixes the mix % near 20–30% produces maximum performance and emissions attributes. The less carbon structure alcohols such as CH_3OH and C_2H_5OH are requires added substance substances because of these alcohols blending in diesel are believed to be unstable, for instance, emulsifier to stabilize the mix [22].

In case of exhaust parameter, however, unmistakable changes of the emanation of NO_x (increment) and hydrocarbon (decline) if there should arise an occurrence of both bio-energies (institutionalized biodiesel and non-institutionalized TBK-biodiesel) can be gotten, which are connected to the moderate improving impact of the burning. In the event that one biofuel is compared with another biofuel, there is slight distinction in regards of the NO_x, CO, and HC and furthermore shows low, however, biased distinction from pure fossil to pure biofuel. NO_x has expanded, yet HC has declined. Be that as it may, there is an improvement in the ignition procedure. Nitrogen oxides increased the most, at 5.3%, and HC by 30.9%. Therefore, the physical and chemical properties of biofuels can be compensated by another reduced emission component, such as CO and HC due to higher oxygen content [23].

Transestrification process is used to produce cotton seed oil methyl ester (CSOME). This is very promising and efficient method. During this process, maximum amount of biofuel produce if 20% CH_3OH blend with the raw material cotton seed biofuel and 55 °C temperature of reaction, 0.5% by weight of catalyst and a reaction time of around 8 h. It was found by the researcher that BSFC are higher and lower thermal efficiency in comparision with diesel fuel due to lower calorific value of the fuel [24, 25].

These perceptions are ascribed to bring down heating value, poor quality of air–fuel blending furthermore, atomization of fuel and higher density, viscosity of the biofuel. Biodiesel combinations indicated lesser smoke, PM [26], and carbon mono oxides emissions; furthermore, more NO_X [27] discharged due to their extra O_2 molecule due to which proper burning of fuel takes place inside the combustion chamber [28, 29].

It was seen by the researcher that BSFC increased for the palm oil biodiesel than the standard biofuel. Palm oil has the lower calorific value in comparision with diesel oil. So that blending of palm oil with diesel decreased the BSFC. Experimentally, it was seen by the researcher that for the biofuel blending ratio like as B25, B50, and B75, these are about to 2.6%, 8.9%, and 9.3% individually. When biofuel mixed with diesel, BTE was found lower due to lower calorific value of the biofuel. This is ascribed to the lower heating value of palm oil. When biofuel mixes with diesel, proper combustion of fuel takes place inside the combustion chamber. Biofuel is more oxygenated fuel. These oxygen molecules support the burning of fuel. So that CO emissions produced due to improper burning of fuel reduced. Experimentally, it was seen that CO emissions particle for the B100, B75, B50, and B25 are 53, 35, 35, and 21%, respectively, with the diesel fuel at full-load conditions [30, 31]. Palm oil is more oxygenated fuel in comparision with diesel oil due to which it provides more oxygen during the combustion of fuel resulting proper combustion occurred inside the combustion chamber. So that CO emissions decrease for palm oil blending with diesel at different loading conditions [32]. Due to more oxygen molecules in comparison to diesel fuel, combustion of fuel occurs at higher temperatures (higher flame temperature) inside the combustion chamber. At higher combustion temperature, oxides of nitrogen formed. Notwithstanding, the palm biodiesel expanded the NO_X outflow contrasted with the diesel fuel [33, 34]. Lower HC outflow and smoke emissions are watched for higher biodiesel mixes (over half) by 38.09 and 19% separately [30, 34]. It was seen that when biofuel blended with the diesel, the cylinder peak pressure increased than the standard diesel fuel. Biofuel has more oxygen molecule in comparision with diesel fuel. So that early ignition occurred inside the combustion chamber. On the another aspect, it has the higher viscosity and density in comparision with diesel, so that during fuel injection it produces higher pressure wave due to which injection valve opens slightly before the set time. Biofuel oil has the short chain unsaturated fatty acid which has the low boiling point in comparision with standard diesel fuel due to which quick burning of biofuel occurred inside the combustion chamber. These are the main reason due to which ignition delay of biofuel reduced in comparision with standard diesel fuel [34].

JOME is a non-edible biofuel. Its physical properties are close to the diesel due to which it is suitable to use with diesel in diesel engine. It was seen that JOME biofuel has the higher thermo-physical properties in comparision to diesel fuel such as viscosity, density, flash point, fire point, and cetane number, but it has the lower calorific value, lower volatility, and high iodine number in comparision with diesel fuel due to the presence of double bound of oxygen molecule in biofuel structure. It was seen that utilization of Jatropha biodiesel decreases the brake thermal efficiency due to lower calorific value of the fuel and also decreased the particulate matter [35–37] and smoke discharges. It was seen by the researcher that ignition delay reduced when biofuel blend with the diesel fuel in different ratio because of biofuel has high density, less compressibility, high viscosity, and unsaturated fatty acids. On the aspect, cylinder peak pressure increases because of biofuel which is an oxygenated fuel. It offers the oxygen molecule in combustion of fuel. So combustion of fuel occurred at high temperature and pressure. Biofuel has the low calorific value and higher density and viscosity due to which blending of biofuel in diesel BSFC increased. Due to biofuel blending in diesel combustion of fuel, flame temperature increased. At higher temperature, oxides of nitrogen formed. NO_X emissions increased by increasing the ratio of biofuel in diesel. Oxides of nitrogen formed at the higher temperature.

It very well may be induced from these trial examinations on Jatropha that there is slight decrement in engine performance, however, with improved emissions attributes at higher compression ratio and higher injection pressures. Notwithstanding, it was found that at all compression ratio and injection pressure, oxides of nitrogen are more in comparision with standard diesel fuel. When jatropha biofuel is preheated, its viscosity and density decreased which supports in proper atomization of fuel inside the cylinder resulting proper combustion occurred. So preheated jatropha biofuel may be the promising fuel at higher compression ratio and injection pressure. As above-discussed, blending of jatropha in diesel reduced the emissions like as CO and HC due to which it may be proved better alternative biodiesel at higher compression and injection pressure [38].

Researcher studied the effect of the preheating of biofuel. The engine performance and emission characteristics are analyzed for the preheated rapeseed oil and diesel on versatile diesel engine. When biofuel is preheated, its viscosity is reduced. Preheated oil influenced engine BSFC, BTE, and emissions characteristics like as CO_2, CO, HC, and NO_X. Compared with preheated oil, BTE is reduced to the not preheated mixture. Not preheated mixture reduces NO_X emissions because of the lower peak combustion temperature [39]. It was observed that without preheated oil reduces the NO_X emissions due to lower peak combustion temperature inside the combustion chamber. However it is found that NO_X emissions of the preheated fuel are higher than other test fuels. Higher NO_X emissions, the reason from the preheated mixture is the increase in combustion temperature. Due to preheating of oil, the inlet fuel temperature increases which reduces the CO and smoke density. On the another aspect, preheated biofuel has the better spray characteristics and air–fuel mixing resulting reduction in CO and smoke capacity [40, 41].

It was found that biofuel has the lower calorific value in comparision with diesel fuel. So that BSFC in decreases and brake power decrease of diesel engine when blend ratio of biofuel increases in diesel fuel. Engine discharges, for example, NOx was higher because of biofuel combustion occurred at higher flame temperature due to oxygenated fuel [42]. Researcher discovered that 5, 10, 20, and 30% jatropha mixes with diesel, the engine emissions like as give BTE, HC, CO, and CO_2 which are lower in comparison with diesel but BSFC and NOx are higher than diesel fuel. So that it was estimated that jatropha biodiesel can be mixed with the diesel oil, and use of blending of jatropha with diesel is compatible as petroleum oil [43]. Researcher investigated the diesel engine performance and emission characteristics when moringa oil methyl ester used as the biofuel with diesel, and it can be mixed in different proportions from B20 to B100 in a DI diesel engine at different burden condition. Researcher analysis the performance of the diesel engine when moringa biofuel blend with diesel fuel used in unmodified diesel engine. It was found that blended fuel have lower brake thermal efficiency due to lower calorific value and higher density of fuel than diesel fuel. On account of engine emissions, moringa mix produces lower HC, CO, and PM emissions; however, higher NOx discharge contrasted with diesel fuel [44].

The study found that purified jatropha tree biodiesel and heated leprosy tree biodiesel produced less braking power on the output shaft than diesel. Nitrogen oxide discharges are higher than diesel because of the significant level of unsaturated fats in jatropha tree biodiesel [45].

The researchers announced that the purified biodiesel comes from a mixture of beef fat and its mixture ratio that reduces BTE and makes BSFC compared to diesel fuel due to the low thermal value of the carnivore biodiesel. I found that emissions like CO and UHC decreased and smoke density decreased by 24.0, 32.45, and 63.0% alone, but increased NOx emissions due to the high oxygen molecules in biodiesel structures [46].

Biodiesel is obtained from the various types of feed stock. As we are probably aware, biodiesel produced from vegetable oil, animal fats, and algae ordinarily comprises of five significant long carbon chain FAMEs. It was seen by the researcher that saturated FAME is in biofuel molecular structure which are methyl palmitate (C16:0) and methyl stearate (C18:0), where first number shows the carbon number and second number show the double bond. If double bond is zero, it means that it is a saturated FAME. But on another aspect, some unsaturated FAME are found in molecular structure are oleate (C18:1), linoleate (C18:2), and linolenate (C18:3). If any unsaturated FAME has the single, double, and triple double bond, they know as the mono-, bi-, and tri-unsaturated FAME [47]. It was seen by the researcher that FAME directly influenced the physical properties like as cetane number, flash point, fire point, ignition delay, viscosity, density, calorific value, iodine value, and also majorly affect on performance and emissions of diesel engine like as CO, CO_2, HC, and NO_X. FAME of the biofuel plays a significance role in the emission characteristics of biofuel. It was seen by the researcher that NO_X emissions increase with the increase of blending ratio of biofuel in standard diesel which means that NO_X emissions are directly concern to double bond or may say that O_2 molecules. Unsaturated FAME gives the more oxygen molecule in combustion of fuel resulting flame temperature of combustion increases and at the higher temperature NO_X formed [48, 49]. So that ignition delay also influenced by the instauration level of biofuel because of ignition delay is defined as the time between the start of injection of fuel and ignition of fuel. Instead of it due to double-bond biofuel has the less compressibility and high density and viscosity, so that biofuel produce pressure wave inside the injector vale due to which it opens slightly before set standard time. Higher saturation level can abbreviate the substance start postpone time, be that as it may, the physical start would be delay with the expansion of C16:0 and C18:0, just as C18:1, on the grounds that their generally higher kinematic viscosity are not great for the fuel atomization, fuel–air blending and evaporation qualities of biodiesel, and decreasing the burning efficiency. Consequently rapeseeds biodiesel created the most reduced showed power and the least NOx emissions.

3 Conclusion

This literature review provides comprehensive information about the combustion and emission characteristics of promising alternative biofuel diesel blending fuel. Biodiesel is sustainable, biodegradable, and eco-friendly fuel. It has the higher flash point, so that it is easy safe to handle, store, and transport. Biodiesel obtained from free fatty acids of vegetable oils and animal which is converted into ester by transesterification process. In compare to diesel, biodiesel emits lower CO, HC and PM and higher CO_2 and oxides of nitrogen. It was observed that biodiesel fuels have higher viscosity, density, cetane number, and lower compressibility in comparision with the diesel fuels due to which biodiesel has an early start of combustion and a shorter ignition delay.

From the systemic review it was concluded that:-

- Transesterifcation process is a promising technique to reduce the viscosity of biofuel.
- The physical and chemical properties of biofuel depend on the fatty acid compositions of extracted oil.
- Biodiesel has the lower calorific value, so that it produces the lower HRR, BTE, and higher BSFC than the standard diesel oil.
- From the quantitative comparisons, it was seen by the researcher biofuel blend produces the leseer HC, CO, and smoke abut NO_X and CO_2 emissions increased in comparision with diesel fuel because of biofuel is more oxygenated fuel in comparision with diesel fuel.
- Biodiesel is without sulpher and aromatic compounds, so that it does not produce any sulfurous emissions component.nat role of better alternative fuel in diesel engine without change in design of diesel engine.

4 Future Scope

It was analyzed from the literature review that there is more work needed on biodiesel to analyze the diesel engine characteristics with biodiesel. It was seen that biodiesel is currently not economical due to non-availability in large quantity. On another aspect, it is efficient to use 20% lend with diesel fuel. So that advanced technology and research are required to enhance the importance of biodiesel with other conventional energy sources. Therefore, more researcher have been conducted on disadvantage of biofuel such as engine deposits, fuel injector clogging, and engine pollution. Researchers are looking that engine must be performed with biofuel by eliminating the weakness of without change in design.

References

1. McCormick, R.L., Williams, A., Ireland, J., Brimhall, M., Hayes, A.A.: Effects of biodiesel blends on vehicle emissions. Milestone report NREL/MP-540-40554. National Renewable Energy Laboratory, Colorado (2006)
2. Canakci, M.: Combustion characteristics of a turbocharged DI compression ignition engine fueled with petroleum diesel fuels and biodiesel. Bioresour. Technol. **98**, 1167e75 (2006)
3. Ma, F., Hanna, M.A.: Biodiesel production: a review. Bioresour. Technol. **70**(1), 1e15 (1999)
4. Encinar, J.M., González, J.F., Rodríguez-Reinares, A.: Ethanolysis of used frying oil. Biodiesel preparation and characterization. Fuel Process. Technol. **88**(5), 513e22 (2007)
5. Meher, L.C., Vidya, S.D., Naik, S.N.: Technical aspects of biodiesel production by transesterification—a review. Renew. Sustain. Energy Rev. **10**(3), 248e68 (2006)
6. Darnoko, D., Cheryan, M.: Kinetics of palm oil transesterfication in a batch reactor. JAOCS **77**(12), 1263e7 (2000)
7. Fukuda, H., Kondo, A., Noda, H.J.: Biodiesel fuel production by transesterification of oils. J. Biosci. Bioeng. **92**(5), 405e16 (2001)
8. Bannister, C., Chuck, D.C., Bounds, M., Hawley, J.G.: Oxidative stability of biodiesel fuel. Proc. IMechE Part D J. Automob. Eng. **225**(1), 1e16 (2010)
9. Knothe, G., Van Gerpen, J., Krahl, J.E.: The biodiesel handbook. AOCS Press, Illinois (2005)
10. Misra, R.D., Murthy, M.S.: Blending of additives with biodiesels to improve the cold flow properties, combustion and emission performance in a compression ignition engine—a review. Renew. Sustain. Energy Rev. **15**, 2413–2422 (2011)
11. Sun, J., Caton, J.A., Timothy, J.: Oxides of nitrogen emissions from biodiesel-fuelled diesel engines. Progr. Energy Combust. Sci. **36**, 677–695 (2010)
12. Ogunkoya, D., Roberts, W.L., Fang, T., Thapaliya, N.: Investigation of the effects of renewable diesel fuels on engine performance, combustion, and emissions. Fuel **140**, 541–554 (2015)
13. Suh, H.K., Lee, C.S.: A review on atomization and exhaust emissions of a biodiesel fueled compression ignition engine. Renew. Sustain. Energy Rev. **58**, 1601–1620 (2016)
14. Yunuskhan, T.M., Atabani, A.E., Badruddin, I.A., Badarudin, A., Khayoon, M.S., Triwahyono, S.: Recent scenario and technologies to utilize non-edible oils for biodiesel production. Renew. Sustain. Energy Rev. **37**, 840–851 (2014)
15. Gautam, R., Kumar, N., Sharma, P.: Comparative assessment of performance emission and combustion characteristics of blends of methyl and ethyl ester of jatropha oil and diesel in compression ignition engine. SAE International 213-01-2664, published 14 Oct 2013
16. Singh, V., Agarwal, T., Saroha, N., Gautam, R.: Performance emissions and combustion analysis of CI engine using ethyl ester kusum oil and butanol blends. SAE Technical Paper 2019-01-0568 (2019). https://doi.org/10.4271/2019-01-0568
17. Gautam, R., Ansari, N.A., Thakur, P., Sharma, A., Singh, Y.: Status of biofuel in india with production and performance characteristics: a review. 1630298 (2019). https://doi.org/10.1080/01430750.2019
18. Dharma, S., Ong, H.C., Masjuki, H.H., Sebayang, A.H., Silitonga, A.S.: An overview of engine durability and compatibility using biodiesel–bioethanol–diesel blends in compression-ignition engines. Energy Convers. Manag. **128**, 66–81 (2016)

19. Nwafor, O.M.I.: Effect of advanced injection timing on emission characteristics of diesel engine running on natural gas. 0960-1481/$—see front matter, Elsevier Ltd., 2007. All rights reserved. https://doi.org/10.1016/j.renene.2006.12.006
20. Savariraj, S., Ganapathy, T., Saravanan, C.G.: Performance, emission and combustion characteristics of fish-oil biodiesel engine. Eur. J. Appl. Eng. Sci. Res. **2**(3), 26–32 (2013)
21. Singh, R.C., Chaudhary, R., Pandey, R.K., Maji, S., Babbar, A., Chauhan, B.S., Gautam, R., Mishra, C.: Performance evaluation of an air cooled diesel engine fuelled with neat neem oil and diesel blend. J. Biofuels **3**, 1–10 (2012)
22. Gautam, R., Ansari, N.A., Thakur, P., Sharma, A., Singh, Y.: Status of biofuel in India with production and performance characteristics: a review. Int. J. Ambient Energy (2019). https://doi.org/10.1080/01430750.2019.1630298
23. Szabados, György, Bereczky, Ákos: Experimental investigation of physicochemical properties of diesel, biodiesel and TBK-biodiesel fuels and combustion and emission analysis in CI internal combustion engine. Renew. Energy (2018). https://doi.org/10.1016/j.renene.2018.01.048
24. Nabi, M.N., Rahman, M.M., Akhter, M.S.: Biodiesel from cotton seed oil and its effect on engine performance and exhaust emissions. Appl. Therm. Eng. **29**, 2265–2270 (2009)
25. Gumus, M., Kasifoglu, S.: Performance and emission evaluation of a compression ignition engine using a biodiesel (apricot seed kernel oil methyl ester) and its blends with diesel fuel. Biomass Bioenergy **34**, 134–139 (2010)
26. Su, F., Peng, C., Li, G.L., Xu, L., Yan, Y.J.: Biodiesel production from woody oil catalyzed by *Candida rugosa* lipase in ionic liquid. Renew. Energy **90**, 329–335 (2016)
27. Hoekman, S.K., Robbins, C.: Review of the effects of biodiesel on NOx emissions. Fuel Process. Technol. **96**, 237–249 (2012)
28. Thangaraja, J., Anand, K., Mehta, P.S.: Biodiesel NOx penalty and control measures—a review. Renew. Sustain. Energy Rev. **61**, 1–24 (2016)
29. Bhuiya, M.M.K., Rasul, M.G., Khan, M.M.K., Ashwath, N., Azad, A.K., Hazrat, M.A.: Prospects of 2nd generation biodiesel as a sustainable fuel—part2: properties, performance and emission characteristics. Renew. Sustain. Energy Rev. **55**, 1129–1146 (2016)
30. Chong, C.T., Ng, J.H., Ahmad, S., Rajoo, S.: Oxygenated palm biodiesel: ignition, combustion and emissions quantification in a light-duty diesel engine. Energy Convers. Manag. **101**, 317–325 (2015)
31. Sharon, H., Karuppasamy, K., Kumar, D.R.S., Sundaresan, A.: A test on DI diesel engine fueled with methyl esters of used palm oil. Renew. Energy **47**, 160–166 (2012)
32. Imtenan, S., Masjuki, H.H., Varman, M., Kalam, M.A., Arbab, M.I., Sajjad, H., et al.: Impact of oxygenated additives to palm and jatropha biodiesel blends in the context of performance and emissions characteristics of a light-duty diesel engine. Energy Convers. Manag. **83**, 149–158 (2014)
33. Rahman, S.M.A., Masjuki, H.H., Kalam, M.A., Abedin, M.J., Sanjid, A.: Rahman MM Assessing idling effects on a compression ignition engine fueled with Jatropha and Palm biodiesel blends. Renew. Energy **68**, 644–650 (2014)
34. Senthilkumar, S., Sivakumar, G., Manoharan, S.: Investigation of palm methyl-ester bio-diesel with additive on performance and emission characteristics of a diesel engine under 8-mode testing cycle. Alex Eng. J. **54**, 423–428 (2015)
35. Paul, G., Datta, A., Mandal, B.K.: An experimental and numerical investigation of the performance, combustion and emission characteristics of a diesel engine fueled with jatropha biodiesel. Energy Proc. **54**, 455–467 (2014)
36. Sangeeta Moka, S., Pande, M., Rani, M., Gakhar, R., Sharma, M., et al.: Alternative fuels: an overview of current trends and scope for future. Renew. Sustain. Energy Rev. **32**, 697–712 (2014)
37. Abedin, M.J., Imran, A., Masjuki, H.H., Kalam, M.A., Shahir, S.A., Varman, M., et al.: An overview on comparative engine performance and emission characteristics of different techniques involved in diesel engine as dual-fuel engine operation. Renew. Sustain. Energy Rev. **60**, 306–316 (2016)

38. No, S.Y.: Inedible vegetable oils and their derivatives for alternative diesel fuels in CI engines: a review. Renew. Sustain. Energy Rev. **15**, 131–149 (2011)
39. Zhong, W., Xuan, T., He, Z., Wang, Q., Li, D., Zhang, X., et al.: Experimental study of combustion and emission characteristics of diesel engine with diesel/second generation biodiesel blending fuels. Energy Convers. Manag. **121**, 241–250 (2016)
40. Buyukkaya, E.: Effects of biodiesel on a DI diesel engine performance, emission and combustion characteristics. Fuel **89**, 3099–3105 (2010)
41. Çelikten, I., Mutlu, E., Solmaz, H.: Variation of performance and emission characteristics of a diesel engine fuelled with diesel, rapeseed oil and hazelnut oil methyl ester blends. Renew. Energy **48**, 122–126 (2012)
42. Altaie, M.A.H., Janius, R.B., Rashid, U., et al.: Performance and exhaust emission characteristics of direct-injection diesel engine fueled with enriched biodiesel. Energy Convers. Manag. **106**, 365–372 (2015)
43. Chauhan, B.S., Kumar, N., Cho, H.M.: A study on the performance and emission of a diesel engine fueled with Jatropha biodiesel oil and its blends. Energy **37**, 616–622 (2012)
44. Rajaraman, S., Yashwanth, G.K., Rajan, T. et al.: Experimental investigations of performance and emission characteristics of Moringa oil methyl ester and its diesel blends in a single cylinder direct injection diesel engine. In: Proceedings of the ASME 2009 International Mechanical Engineering Congress & Exposition, 13–19 November, Lake Buena Vista, Florida, USA, 2009
45. Rao, P.V.: Experimental investigations on the influence of properties of jatropha biodiesel on performance, combustion, and emission characteristics of a DI-CI engine. World Acad. Sci. Eng. Technol. **75**, 855–868 (2011)
46. Selvam, D.J.P., Vadivel, K.: Performance and emission analysis of DI diesel engine fuelled with methyl esters of beef tallow and diesel blends. Proc. Eng. **38**, 342–358 (2012)
47. Selvan, T., Nagarajan, G.: Combustion and emission characteristics of a diesel engine fuelled with biodiesel having varying saturated fatty acid composition. Int. J. Green Energy **10**(9), 952–965 (2013)
48. Ozsezen, A.N., Canakci, M.: Determination of performance and combustion characteristics of a diesel engine fueled with canola and waste palm oil methyl. publ.-Energy Convers. Manage. **52**(1), 108–116 (2011)
49. Herbinet, O., Pitz, W.J., Westbrook, C.K.: Detailed chemical kinetic oxidation mechanism for a biodiesel surrogate. Combust. Flame **154**(3), 507–528 (2008)

Analysis of Different Techniques of Superconductivity

Pooja Rani

1 Introduction

Superconductivity is a very interesting topic that applicable in our daily life experiences. Critical temperature is the temperature when electric resistivity becomes zero. Superconductivity was discovered by Heike Onnes Kamerling in 1911. A superconducting material changes into normal conductor when its superconductivity vanishes [1].

1.1 Meissner Effect

In 1933, Walther Meissner and Robert Oschsenfeld validate that superconductors provide altered material properties than ideal material. They discovered that magnetic field lines of material (Lead and Tin) do not enter the material which gives zero magnetic fields inside the material called Meissner effect [1] as shown in Fig. 1. From this finding, we found that a superconductor has a diamagnetic property. But ideal conductor was not exhibited diamagnetic property although zero resistivity. From this, it was found that a new type of matter with certain property [2].

In superconductor, magnetic field is zero it means $B = 0$.

$$\begin{aligned}
&\mu_o(H+M)=0\\
&H+M=0\\
&M=-H\\
&\tfrac{M}{H}=-1\\
&\text{But}\,\chi=M/H\\
&\text{So},\chi=-1
\end{aligned}$$

P. Rani (✉)
G L Bajaj Institute of Technology and Management, Greater Noida, UP, India

L. M. Das et al. (eds.), *Recent Trends in Thermal Engineering*,
Lecture Notes in Mechanical Engineering,
https://doi.org/10.1007/978-981-16-3428-4_4

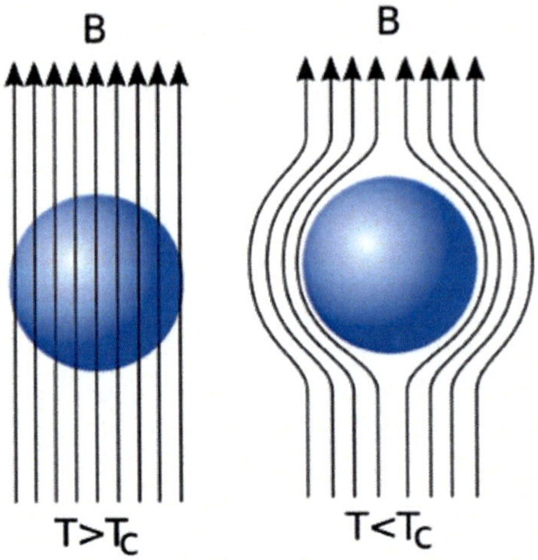

Fig. 1 Magnetic field lines of material (lead and tin) do not enter the material which gives zero magnetic fields inside the material called Meissner's effect

Susceptibility is negative; it means material is diamagnetic [3].

1.2 Silsbee Rule

When the current has more value than critical value then it loses superconducting state called Silsbee rule [4].

$$I_c = 2\pi r H_c$$

1.3 Classification of Superconductors

The superconductors may be classified into two kinds like Type I and Type II superconductors.

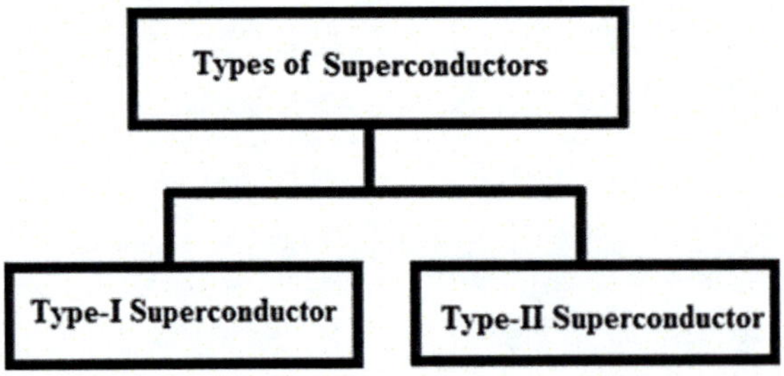

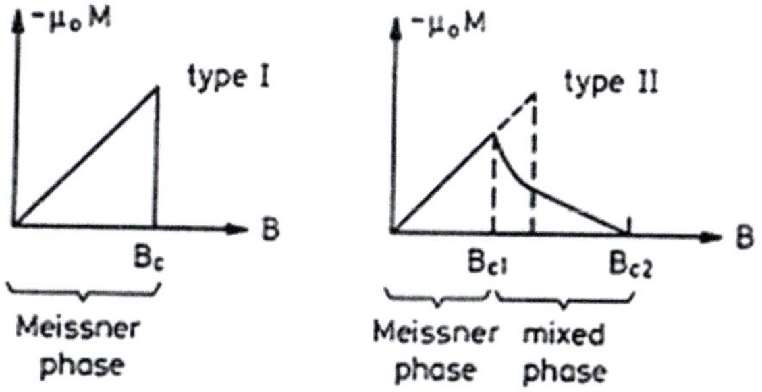

Fig. 2 Type I and type II superconductors

Type I superconductors are metallic alloys and metals. The critical temperature lowers when current passes through the superconductor. The critical temperature decreases with an application of external magnetic field. The different effects are found on the material when magnetic field becomes more as compared to critical value [5]. Other name of Type I superconductors are soft superconductors or low-temperature superconductors. At very low temperature, it obtains superconductive properties. It has no mixed state. It follows Meissner effect [6].

On the other hand, superconductors type II or high-temperature superconductors have two critical magnetic fields. When the field starts to enter, the conductor is designed as lower field and the superconductivity breaks down in the higher field. The area between lower and upper critical field is called mixed state. Other name of type II superconductor is hard superconductor. It does not follow Meissner effect [2]. Type I and Type II superconductors are shown in Fig. 2.

1.4 BCS Theory

Bardeen, Cooper and Schrieffer (BCS) suggested a BCS theory in 1957. It described the origin of superconductivity and gives information about properties of superconductor material. It is very difficult theory. In the absence of electrical resistance, the ability of certain metal is to conduct electricity at low temperature [7].

a. The forced oscillations of an ion core caused by interface with an electron effect the movement of other electron transient through nearby it and give rise to an attraction between two electrons [8].
b. Cooper pairs are formed when the two electrons which interact attractively with each other in phonon fields shown in Fig. 3.
c. When T greater than T_C, electron–electron coulomb contact is more than electron-lattice-electron interaction.
d. When T less than T_C, electron-lattice-electron coulomb contact is more than electron–electron coulomb contact, and so electrons tend to pair up [9].
e. Pairing thus initializes when temperature is low as compared to the critical temperature.

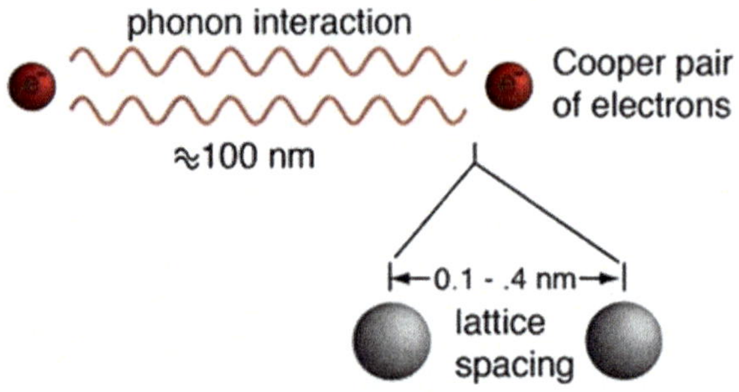

Fig. 3 Formation of Cooper pair

f. Cooper pairs may not be distributed because flat riding on the lattice deficiencies lacking any interchange of energy with them and hence show infinite conductivity [10].
g. The Cooper pairs may sustain their together movement up to a definite distance, called coherence length [11].

1.5 High-Temperature Superconductors

The materials act like superconductors at temperature more than 77 K called high-temperature superconductors. HTS may be cooled by liquid nitrogen. The HTSC may be chilled larger powerfully [12]. In SQUIDS, small variation of magnetic field may be measured. High-temperature SQUID has more noise as compared to older helium technology, but accepted due to cost and profits of nitrogen cooling [13].

1.6 Applications of Superconductors

Superconductors can be used in the many applications like:

a. Maglev (magnetic levitation) trains. In this application, a magnet can drift on a superconductor because a superconductor repels a magnetic field. By this friction diminish between the train and the track. Safety measures are required for high magnetic field to human health [14].
b. They are used to make electromagnets to go faster-charged particles.
c. SQUIDs (superconducting quantum interference devices) may be used to discover the weakest magnetic field [15].
d. Electricity may be generated more resourceful.
e. They are used in magnetic resonance imaging (MRI) and nuclear magnetic resonance (NMR) machines [12].

1.7 Recent Research Paper Work in Superconductivity

a. Misfit phase (BiSe) $_{1.10}$ $NbSe_2$ as the origin of superconductivity in niobium-doped bismuth selenide.
b. Room-temperature superconductivity in a carbonaceous sulfur hydride.
c. Ultrafast transient liquid assisted growth of high current density superconducting films.
d Unique defect structure and advantageous vortex pinning properties in superconducting $CaKFe_4As_4$.
e. Superconductivity in a 122-type Fe-based compound (La, Na, K) Fe_2As_2.
f. Fast Na ion transport triggered by rapid ion exchange on local length scales.
g. Fabrication of 3D Fingerprint Phantoms via Unconventional Polycarbonate Molding.
h. Tracking aluminum impurities in single crystals of the heavy-fermion superconductor UBe_{13}.
i. Highly reproducible alkali metal doping system for organic crystals through enhanced diffusion of alkali metal by secondary thermal activation.

2 Result Analysis: Important Parameters Corresponding to Superconductivity Are Described Below

a. **Meissner's effect**: Magnetic field lines of material (Lead and Tin) do not enter the material which gives zero magnetic fields inside the material.
b. Silsbee rule: $I_c = 2\pi r H_c$
c. Superconductors can be classified into two categories, i.e., Type I and Type II superconductors
d. BCS theory explains the formation of Cooper pair.
e. High-temperature superconductors are classified as the materials whose temperature more than 77 K.
f. Applications of superconductors: Maglev trains, electromagnets, SQUIDS, MRI, NMR.
g. Recent work in superconductivity: 3D Fingerprints, Tracking aluminum impurities, Superconducting films.

3 Conclusion

In this paper, we provide information about different factors of superconductivity like different types of superconductors, Meissner's effect, Silsbee's rule, BCS theory, applications of superconductivity and discuss various recent researches on the basis of superconductivity. It also tells about use of superconductivity in our daily life experiences. Superconductivity improvises the working efficiency and capability and acts as an 'energy superhighway.' The economic and energy impacts of superconductors are predicted to be huge. Although superconductors are not new technologies, but still there is room of improvement made in research and innovation.

References

1. Meissner, W., Oschsenfeld, R.: Ein neuer Effect bei Eintritt der Supraleitfahigkeit. Naturwissenschaften **21**(44), 787–788 (1933)
2. Branício, P.: Introdução à supercondutividade, Suas Aplicações e Mini-Revolução Provocada Pela Redescoberta do MgB2: Uma abordagem Didática, Revista Brasileira de Ensino de Física, **4**(23) (2001)
3. Abricosov, A.: On the magnetic proprieties of superconductors of the second group. Sov. Phys. JETP **5**(6) (1957)
4. Nishijima, S., et al.: Superconductivity and the environment: a Roadmap. Supercond. Sci. Technol. **26**, 35 (2013)
5. Thomas, H., et al.: Superconducting Transmission Lines—Sustainable Electric Energy Transfer with Higher Public Acceptance. Elsevier (2016)
6. Zeguer, H.: 630 kVA High Temperature Superconducting Transformer, vol. 38(11), pp. 1169–1172. Elsevier (1998)
7. Roque, A., Sousa, D.M., Margato, E., Machado, V., Sebastião, P., Marques, G.: Magnetic flux density distribution 3D analysis in the air gap of a ferromagnetic core with superconducting blocks. IEEE Trans. Appl. Supercond. **25**(6) (2015)
8. Sousa, D.M., Marques, G.: Study of the air gap magnetic field distribution of a nuclear magnetic resonance iron-core magnet. In: 1st IEEE Region 8 International Conference on Computational Technologies in Electrical and Electronics Engineering, Novosibirsk, Russia, pp. 242–247 (2008)
9. Tinkham, M., Lobb, C.J.: Physical properties of the new superconductors. Solid State Phys. **42**, 91 (1988)
10. Noack, F.: NMR field-cycling spectroscopy: principles and aplications. Prog. NMR Spectro. **18**, 171–276 (1986)
11. Ali, M., et al.: An overview of SMES applications in power and energy systems. IEEE Trans. Sustain. Energy **1**(1) (2010)
12. Strasik, M., et al.: Design, fabrication, and test of a 5-kWh/100 kW flywheel energy storage utilizing a high-temperature superconducting bearing. IEEE Trans. Appl. Supercond. **17**(2) (2007)
13. Narlikar, A.: High Temperature Superconductivity 2—Engineering Applications. Springer (2004)
14. Tixador, P., et al.: Design and first tests of a 800 kJ HTS SMES. IEEE Trans. Appl. Supercond. **17**(2) (2007)
15. Xue, X., Cheng, K.D, Sutanto, D.: Power system applications of superconducting magnetic energy storage systems. In: Fourtieth IAS Annual Meeting (2005)

Prediction of the Dynamic Viscosity of MXene/palm Oil Nanofluid Using Support Vector Regression

Naman Parashar, Junaid Khan, Navid Aslfattahi, R. Saidur, and Syed Mohd Yahya

1 Introduction

Nanofluid, first developed by Choi and Eastmen [1], is a class of fluid where nanometer sized particles (upto size of 100 nm) are suspended in fluids with the aim of enhancing their thermophysical properties. These nanometer-sized particles are known as nanoparticles and the fluids in which they are suspended are known as base fluids. Various nanoparticles that are being used include metals, metallic oxides, carbides, ceramics, carbon nanotubes, etc. [2]. Various fluids that are used as base fluids include water, ethylene glycol, water ethylene glycol mixture, oils, etc. It is due to the enhanced thermophysical properties of the nanofluid that they have the potential of being used in a wide range of industries [3]. Various parameters such as nanoparticles diameter, nanoparticles concentration, temperature are known to affect these thermophysical properties. However, with the addition of nanoparticles in the base fluids, not only the properties such as thermal conductivity, heat transfer rate improves, but viscosity also increases. This increase in viscosity is not a desired property as it increases the required pump power [4]. Therefore, it is very important to know viscosity and other properties before a nanofluid is put into application. Evidently, experimental determination of viscosity involves a considerable amount of time, money and energy. Therefore, researchers around the world are finding different ways so that the need of finding the viscosity experimentally can be eliminated or even reduced. One such way is by creating correlations using experimental data. These correlations once developed, may be

N. Parashar · J. Khan (✉) · S. M. Yahya
Sustainable Energy and Acoustics Research Laboratory, Department of Mechanical Engineering, Aligarh Muslim University, Aligarh, India

N. Aslfattahi
Department of Mechanical Engineering, University of Malaya, Kuala Lampur, Malaysia

R. Saidur
School of Science and Technology, Sunway University, Selangor, Malaysia

L. M. Das et al. (eds.), *Recent Trends in Thermal Engineering*,
Lecture Notes in Mechanical Engineering,
https://doi.org/10.1007/978-981-16-3428-4_5

used to determine the values of viscosity at varying ranges of parameters. To achieve this, various researchers developed correlations for different base fluids and nanoparticles. However, one of major issues with these correlations was their accuracy. Therefore, researchers started finding new and more effective ways of modeling the viscosity values of different nanofluids. One way was to use more advanced mathematical models such as artificial neural networks (ANN). Artificial neural networks are the computational models which are a proven tool of learning the mapping between the input values and output values. It is due to the development of algorithms that these models use, ANNs are now widely used in different fields of science and technology. A lot of ANN models were developed by researchers around the world which can be used to predict the thermophysical and rheological properties of nanofluids. These ANN models were more accurate than the correlations [5] in predicting the values of the viscosity. This owes to the more complex algorithms that these ANN models employ. However, these ANN models also have some drawbacks. These drawbacks include [6]: large dataset required to train the model, overfitted models which are not able to give accurate predictions on new data, and prediction uncertainties due to the stochastic nature of ANN models. The drawbacks of the ANN models can be overcome by using other mathematical models such as support vector regression models. Support vector regression (SVR) models are known to have shown better performance even when the data is low. These models are also less prone to overfitting problems and are easy to train because of the less number of parameters involved. Some studies even concluded that the performance shown by the SVR models is better than the performance shown by the ANN models [7, 8]. Surprisingly, the number of studies in which SVR models were used to predict the viscosity of nanofluids are very less in comparison to the studies in which ANN models were used. An SVR model is created in this study to predict the values of dynamic viscosity and the experimental values are used to facilitate the same. Mean absolute error and coefficient of determination are used to determine the efficacy of the developed model. These aforementioned parameters illustrate that the SVR model is able to anticipate the dynamic viscosity values of MXene/palm oil with appreciable precision.

2 Experimental Data

The fluctuations of the dynamic viscosities with respect to temperature at different concentrations are shown in Fig. 1 [9]. The decrement and increment of dynamic viscosity values with respect to rise in temperature and Mxene dosing are clearly observable from figure. At high temperature of nearly 50–55 °C, the increase in Mxene dosing starts to have little to no effect. The helps in ascertaining the fact that increase in nanoparticle dosing at elevated temperature does not have any noticeable effect on dynamic viscosity and therefore does not necessitate arrangements for enhanced pumping power.

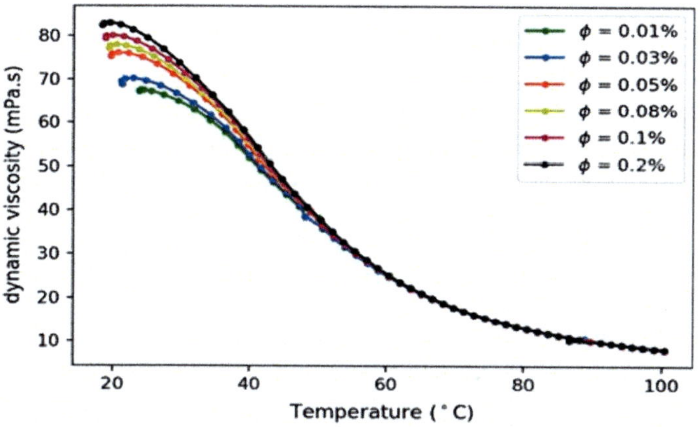

Fig. 1 Variation of viscosity with temperature for different concentrations [9]

3 Support Vector Regression

Support vector regression (SVR) model is a type of support vector machine (SVM) model. Support vector machines models are widely used for classification analysis, whereas support vector regression models are used for regression analysis. The basis of SVR is the statistical learning theory, and it was developed for this purpose [10]. It works by mapping the input data into higher dimensions and then using kernels to give output [11]. These kernels are of different types which include: linear, polynomial, sigmoid, radial basis function (RBF) or Gaussian kernel. Generally, when the relationship between the input variables and output variables is not known, RBF kernel is used. Equation (1) shows the RBF kernel:

$$K(x_i, x_j) = \exp\left(-\gamma \|x_i - x_j\|^2\right) \tag{1}$$

In this study, epsilon (ε)-insensitive type of SVR model was used. Equation (2) shows the regression function that the SVR models have to obtain:

$$f(x) = \sum (\alpha_i^* - \alpha_j) \cdot K(x_i x_j) + \theta \tag{2}$$

where α_i^*, α_i are the lagrange multipliers, $K(x_i, x_j)$ is the kernel function, and θ is the intercept term.

Developing an SVR model includes finding the optimum values of hyperparameters involved in it. These hyperparameters for epsilon insensitive models when RBF kernel is used, are regularization parameter (C), epsilon (ε), and gamma (γ).

Regularization parameter is a trade-off between the training error and flatness of the solution. Increasing the value of C will reduce the training error upto a certain point and after that if the value of C is increased, the model will become more complex and it will give poor performance on new data. Epsilon defines a range upto which no penalty is given to errors. A high value of epsilon will introduce large errors in the solution, whereas a very low value of epsilon will penalize more solution that will result into an over fit model. Hyperparameter gamma is related to RBF kernel. A high value of gamma will result into a model with high bias and low variance, whereas a low value will result into a model with low bias and high variance. Both these type of models are not acceptable. Hence, to develop an optimum model, it is very important to carefully select these hyperparameters.

The entire data was first arbitrarily divided into two sections, respectively, training data and validation data, in ways to construct the model. To create a model, training data was used, while validation data was not fed into the model. This was done to test the efficiency of the model on fresh results. 80% of the data was categorized as training data, i.e., 240 data points, while the remainder of the data was designated as validation data. Concentration and temperature were the input parameters for the model, while dynamic viscosity was the model's performance. The next move was to normalize the information in order to improve the capacity of the model to manage the data.

$$x_{\text{norm}} = (X - \mu)/s \tag{3}$$

where X_{norm} represents the normalized value, μ represents the mean of the variable (s), and s represents the standard deviation of the variable(s).

Next, the optimum set of regularization factor (C), epsilon (ε), and gamma (γ) was found through grid search technique. The grid search technique is widely used to determine the most optimum values of hyperparameters [12]. The optimum values of C, epsilon, and gamma obtained through grid search were 290, 0.0095, and 0.71, respectively.

Performance of the model for training data and validation data was measured using statistical parameters such as coefficient of determination (R^2) and mean absolute error (MAE), which has been presented in the results and discussion section.

4 Results and Discussions

To develop an SVR model, input parameters fed were the concentration and temperature, whereas dynamic viscosity was the output. The optimum values of C, epsilon, and gamma were determined using grid search technique. The SVR model, when developed for these values of parameters, showed the best performance. The values of coefficient of determination (R^2) and mean absolute error (MAE) obtained for training data and validation data were 0.99991 and 0.0070, and 0.999987 and 0.0079, respectively. From these values of the parameters, it can be said that the SVR model is showing good performance on training data as well on validation data.

Figure 2 illustrates the SVR model's efficiency on the training results by comparing the experimental parameters with the expected values of the SVR model. The figure shows a line of "equality" bent similarly to the axes of both. The values that are near the line are more precise than the values that are a little further away from the line. Figure 3 demonstrates the SVR model's success on the validation results. It can be shown from figure that the experimental values are very similar to all the values expected by the SVR model. It also reveals that the SVR model's output on new data is almost as strong as on training data.

The performance of the SVR model has also been described in terms of deviation between the experimentally determined values and SVR predicted values. Equation (4) shows the equation using which deviation was calculated.

$$D = \left((Y_{\text{exp}} - Y_{\text{pred}})/Y_{\text{exp}}\right) * 100 \tag{4}$$

where Y_{exp} represents the experimentally determined values, and Y_{pred} represents the SVR model predicted values.

Figure 4 shows the deviation between experimentally determined values and SVR model predicted values. Maximum deviation between the experimentally determined value and SVR model predicted value is −5.18% as depicted in Fig. 4.

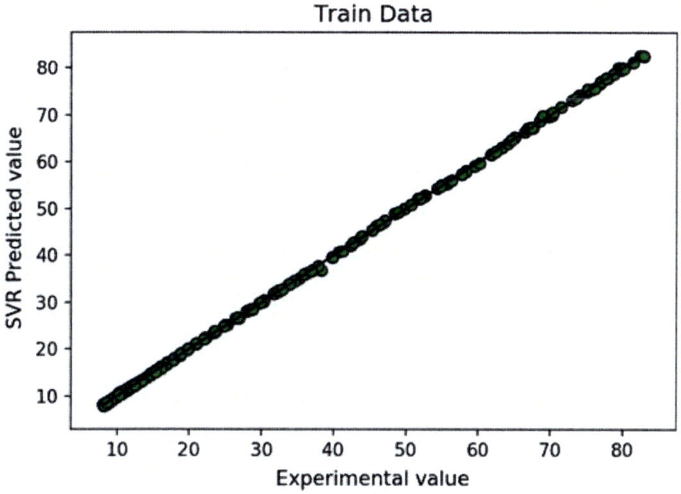

Fig. 2 Performance of the SVR model on the training data

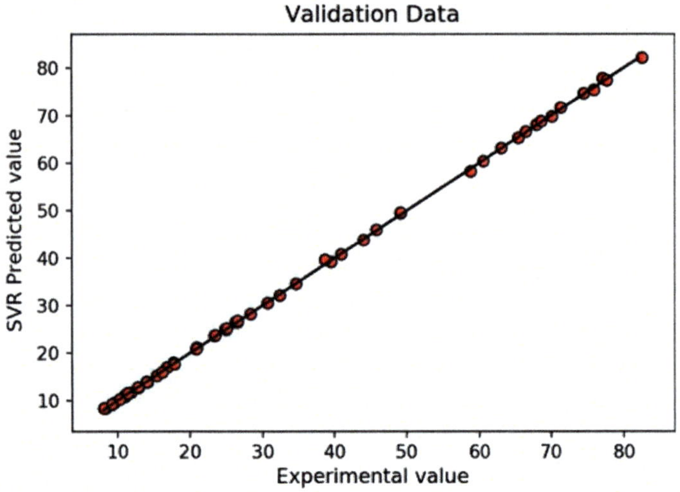

Fig. 3 Performance of the SVR model on the validation data

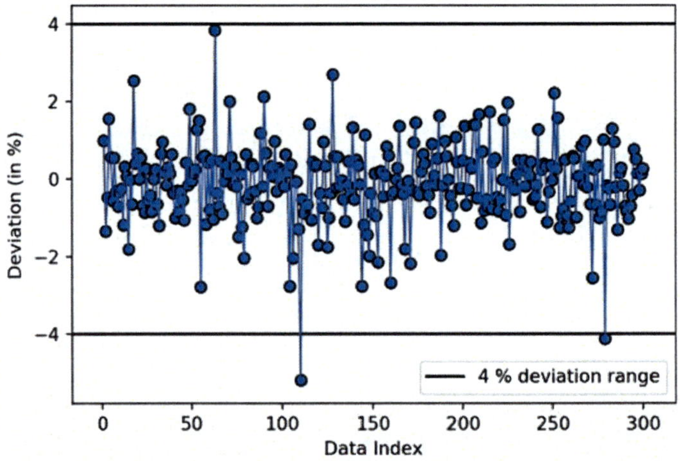

Fig. 4 Deviation between experimental values and SVR predictions

It can also be observed that 99.33% of the deviations are in the range of ±4% and maximum number of points are in the ±2% range.

From the values of statistical parameters shown by the SVR model, it can be said that for predicting the thermophysical properties of nanofluids, SVR models can be good choice in addition to artificial neural networks.

5 Conclusions

The aim of the current study was to demonstrate the SVR models potential and efficacy to estimate the thermophysical properties of nanofluids. Upon utilizing experimentally measured MXene/palm oil nanofluid viscosity values, an SVR model was constructed. Through the grid search technique, the optimum hyper-parameter values were obtained. The correlation coefficient and mean average error values for the test results were found to be 0.99999987 and 0.0079, respectively. The maximum deviation was merely −5.18% between experimentally established values and expected SVR values. 99.33% of deviation points too were found to be in the range of ±4%. Based on the SVR model performance, it can be concluded that SVR models can be a good alternative to artificial neural network in predicting the thermophysical properties of nanofluids.

References

1. Choi, S.U.S.: Enhancing thermal conductivity of fluids with nanoparticles. American Society of Mechanical Engineers, Fluids Engineering Division (Publication) FED, vol. 231, pp. 99–105 (1995)
2. Sajid, M.U., Ali, H.M.: Thermal conductivity of hybrid nanofluids: a critical review. Int. J. Heat Mass Transf. **126**, 211–234 (2018)
3. Yu, W., Xie, H,: A review on nanofluids: preparation, stability mechanisms, and applications. J. Nanomater. **2012**, 17 p (2011)
4. Yahya, S.M., Asjad, M., Khan, Z.A.: Multi-response optimization of TiO_2/EG-water nano-coolant using entropy based preference indexed value (PIV) method. Mater. Res. Express **6** (2019)
5. Ramezanizadeh, M., Ahmadi, M.H., Nazari, M.A., Sadeghzadeh, M., Chen, L.: A review on the utilized machine learning approaches for modeling the dynamic viscosity of nanofluids. Renew. Sustain. Energy Rev. **114**, 109345 (2019)
6. Karimipour, A., Bagherzadeh, S.A., Taghipour, A., Abdollahi, A., Safaei, M.R.: A novel nonlinear regression model of SVR as a substitute for ANN to predict conductivity of MWCNT-CuO/water hybrid nanofluid based on empirical data. Phys. A **521**, 89–97 (2019)
7. Wong, W.-T., Hsu, S.-H.: Application of SVM and ANN for image retrieval. Eur. J. Oper. Res. **173**, 938–950 (2006)
8. Byvatov, E., Fechner, U., Sadowski, J., Schneider, G.: Comparison of support vector machine and artificial neural network systems for drug/nondrug classification. J. Chem. Inf. Comput. Sci. **43**, 1882–1889 (2003)
9. Parashar, N., Aslfattahi, N., Yahya, S.M., Saidur, R.: An artificial neural network approach for the prediction of dynamic viscosity of MXene-palm oil nanofluid using experimental data. J. Therm. Anal. Calorim. (2020). https://doi.org/10.1007/s10973-020-09638-3
10. Vapnik, V.: The Nature of Statistical Learning Theory. Springer, New york (1995)
11. Alade, I.O., Oyehan, T.A., Popoola, I.K., Olatunji, S.O., Aliyu, B.: Modeling thermal conductivity enhancement of metal and metallic oxide nanofluids using support vector regression. Adv. Powder Technol. **29**, 157–167 (2018)
12. Fayed, H.A., Atiya, A.F.: Speed up grid-search for parameter selection of support vector machines. Appl. Soft Comput. **80**, 202–210 (2019)

Experimental and Computational Investigation of Coefficient of Discharge of Venturimeter

Omprakash Yadav, Abhay Dahiya, Vinod Kumar Yadav, and Rahul Sharma

1 Introduction

Venturimeter, a mechanical device, is a short length of pipe with converging, throat and diverging section. Although a number of flows measuring devices like orifice plate, rota meter are available, but venturimeter are more accurate and precise. On the other hand, design and fabrication of venturimeter are very critical. Converging section of venturi is a short length of pipe in which cross sectional area decreases in the direction of flow. Bansal [1] and Daugherty et al. [2] reported that 12° to 20° of convergence angle is a better choice for the design of venturimeter. Throat of the venturimeter is the section with minimum cross sectional area. Diverging section of the venturimeter is larger compared to converging section in which the cross sectional area increases in the direction of flow. Nithin et al. [3], Bansal [1] and Daugherty et al. [2] reported that 4°–7° of convergence angle is a better choice for the design of venturimeter. The divergence angle should be as small as possible [1, 2]. Venturimeter works on the Bernoulli's principle. According to this principle, summation of pressure head, kinetic head and datum head remains same throughout the pipe, if loss of head is zero. To find the theoretical discharge the venturimeter, let us assume two piezometers are inserted at Section-1 and Section 2 to measure the pressure head. Let pressure head difference between two sections is h meter of flowing fluid. In the analysis, P is the pressure, V is the velocity, z is the datum head, ρ is the density of flowing fluid and g is the acceleration due to gravity. Theoretical discharge from the venturimeter can be found by combining Bernoulli and continuity principle.

O. Yadav (✉) · A. Dahiya · V. K. Yadav · R. Sharma
Department of Mechanical Engineering, G.L Bajaj Institute of Technology and Management, Greater Noida, India

L. M. Das et al. (eds.), *Recent Trends in Thermal Engineering*,
Lecture Notes in Mechanical Engineering,
https://doi.org/10.1007/978-981-16-3428-4_6

$$Q_{\text{theoretical}} = A_1.A_2.\sqrt{\frac{2gh}{(A_1^2 - A_2^2)}} \quad (1)$$

Now if U-tube manometer is used, pressure head difference between two sections can be found as

$$h = x.\left[\frac{\rho_{hg}}{\rho} - 1\right] \quad (2)$$

By knowing the actual discharge from experimentation, coefficient of discharge can be found by

$$C_d = \frac{Q_{\text{actual}}}{Q_{\text{theoretical}}} \quad (3)$$

Based upon the literature survey, suitable value of throat diameter, convergence angle, divergence angle and other dimension has been finalized for the inlet diameter of 27 mm. The venturimeter is designed in three parts for the ease of manufacturing. An optimized design of venturimeter by SolidWorks 2016 is shown in Fig. 1.

2 Literature Survey

Literature reviews provide important information of the research which has been done in past and current status of any research work.

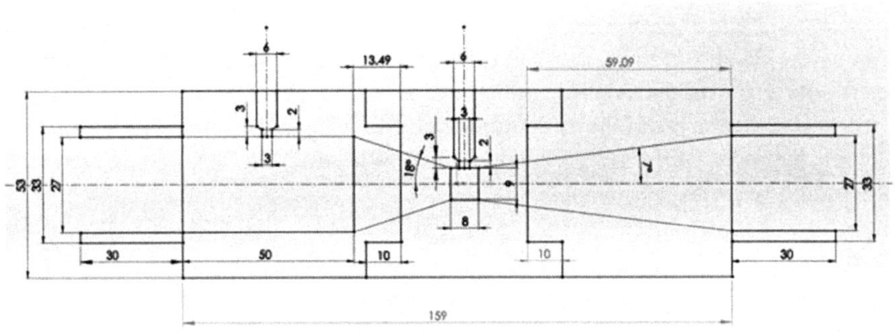

Fig. 1 Design and dimension of venturimeter

Nithin et al. [3] analyses the effect of divergence angle on coefficient of discharge of venturimeter by computational fluid dynamics method. Conclusion of this work is that small length of throat and smaller angle of divergence gave the minimum pressure loss for maximum difference of pressure. From the computational analysis, it has been found that pressure drop is minimum at certain angle of divergence for a given angle of convergence. Suryawanshi et al. [4] study deals with the improvement in coefficient of discharge and minimizing the losses due to friction in the venturi section. This research work presents the experimental and computational analysis for various diameter ratios (*d*/*D*) and compares the experimental result with computational results. Comparison has been done by plotting various velocity streamlines, contours of pressure across the section and noting the values at various sections, i.e., inlet, throat, etc. It was concluded that diameter ratio (*d*/*D*) of 0.65 is best for maximizing the coefficient of discharge, considering the size of the venturimeter and type of fluid as water. Better geometry improves the flow property. The coefficient of discharge increases with increase in Reynolds number. Sanghani et al. [5] has been investigated the effect of geometrical parameters on pressure drop through the venturimeter section by computational analysis for the flow of water. The study was conducted by varying one parameter and keeping the other parameters constant at a time. The variation was made in diameter ratio (*d*/*D*), convergent angle, divergent angle and throat length. The velocity vector plot and pressure drop were used for result analysis. The results showed that the increase in the convergent angle causes fluctuations in the pressure drop. Increase in convergence angle, decreases the diameter ratio which increases the pressure drop.

Jay Kumar et al. [6] has examined the effect of nozzle angle of venturi section on pressure drop computational analysis on fluent. From this computational analysis, it was found that the pressures at the throat of the venturi decrease with increase in opening of the throttle plate due to more flow of air. Pressure at the venturi section was analyzed for the discharge nozzle angle of 30°, 40° and 45° by computational analysis. From this study, it was concluded that for venturi nozzle angle of 30°, pressure variation through the carburetor section is very uniform, which ensures better atomization and vaporization of the fuel inside the carburetor body. But for other nozzle angle of 40° and 45°, pressure distribution is non-uniform inside the venturi section. Arun et al. [7] analytically and computationally investigated the coefficient of discharge of venturimeter at low Reynolds number. The venturimeter was modeled as per dimensions specified. At the inlet, fully developed laminar velocity profile was prescribed. From this study, it was observed, as Reynolds number decreases, coefficient of discharge of venturimeter reduces very fast. This was happening due to the viscous losses on account of low Reynolds number. While using venturimeter at low flow rate, it is essential to use the modified value of Cd. That is why use of venturimeter is restricted for low flow rate. Zhang et al. [8] investigated the effect of the geometrical dimension on fluid flow variation and pressure variation in the venturimeter. They also study effect of convergence angle, divergence angle, throat length to diameter ratio and the differential pressure between the two sections of venturimeter by fluent flow analysis solver. As the convergence angle increases, vacuum pressure and mass flow rate decrease. As the divergence angle increases, vacuum pressure and mass flux decrease.

Yadav et al. [9] experimentally investigates the discharge through nozzle by particle image velocimetry system which is an optical flow measuring device. Seeding particles are mixed with fluid before experimental process. The density of seeding particle should be approximately equal to density of fluid. These seeding particles are illuminated by laser and at the same time image of the flow are taken. The coefficient of discharge was found from PIV and U-tube both. This result was also validated by measuring the flow rate by U-tube manometer. It was found that coefficient of discharge is approximately 0.9873 from both the analysis. Tukimin et al. [10] investigated the flow through venturi tube using SKE turbulence model, and results are used to determine values of coefficient of discharge Cd from the three tests performed. Pressure drop and discharge velocity are analyzed by computational fluid dynamics. Values of Cd obtained are 0.97551, 0.98850 and 0.98902 with a percentage error of 0.937% in experimental and numerical results indicating the agreement of SKE turbulence model with experimental data. Many researchers [11–16] adopted experimental and computational techniques to validate their observations.

3 Experimental Setup and Procedure

3.1 Experimental Setup

A converging-diverging type venturimeter with the experimental setup is shown in Fig. 3. A 0.5hp self-priming hydraulic pump is used to pumps the water from inlet tank to discharge measuring container through the venturimeter. The inlet of the pump is connected to the inlet tank through a flexible pipe, and outlet of the pump is connected to the white pvc pipe again by flexible pipe. A pump can be put on/off by the switch on board. A ball type valve is connected just after the pump outlet to control the flow rate through the venturimeter. Two pressure gauges: pressure gauge-1 and pressure gauge-2 are connected just before and after the venturimeter to measure the pressure at inlet and exit of venturimeter. A U-tube manometer, in which mercury is manometric fluid, is used to measure difference of pressure between two sections of venturimeter. By knowing the pressure difference between Section-1 and Section-2, theoretical discharge through the pipe can be found. Exit of the pipe is connected to discharge measuring container which has inbuilt scale. By knowing the volume of water collected in measuring tank in certain time, actual discharge can be found. Hence, coefficient of discharge, i.e., ratio of actual discharge to theoretical discharge can be found. The experimental setup is shown in Fig. 2. A U-tube manometer has been developed in the workshop and used in this experimental setup to measure the pressure difference between the inlet and throat section of venturimeter.

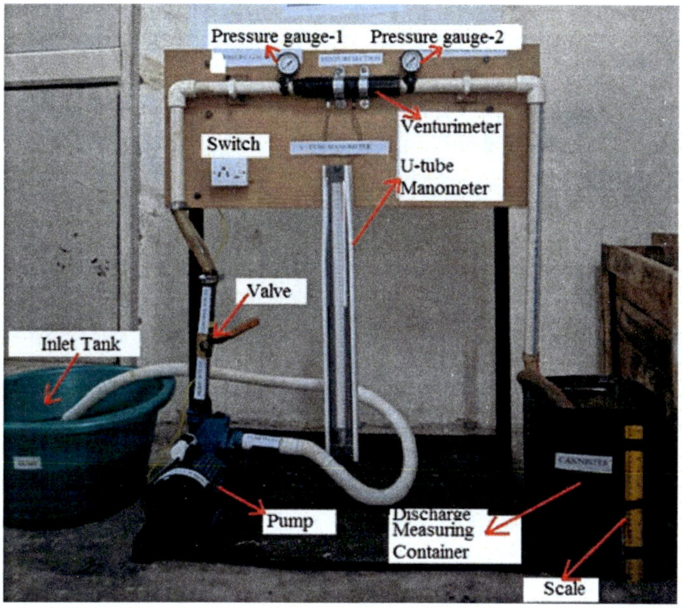

Fig. 2 Experimental setup

Fig. 3 Assembled sectional view of venturimeter

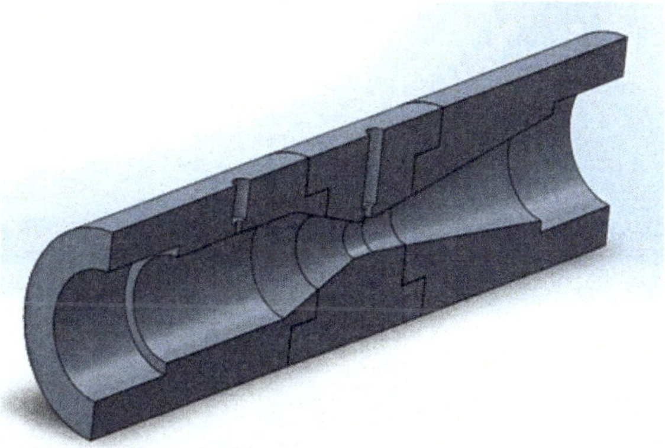

3.2 *Venturimeter*

Venturimeter is made in three parts for the ease of manufacturing and then assembled to each other (Fig. 3). It consists of converging section, throat section and diverging section. The inlet diameter of the pipe is 27 mm, and hence, inlet diameter of the venturimeter was also taken as 27 mm to avoid any kind of jump in the flow. The throat diameter should be 1/3 to 1/4 of inlet diameter [1, 2]. Hence, throat diameter was taken as 9 mm. The convergence angle and length of convergence section are 180 and 27.7 mm, respectively. The divergence angle and length of divergence section are 70 and 59.09 mm, respectively. The length of throat section is 8 mm as shown in Fig. 1. Total length of venturimeter is 20.48 cm out of which 14.48 cm is visible to the eye.

3.3 Discharge Measuring Container with Scale

A container of cross section 25 cm × 25 cm and height 35 cm was used as discharge measuring container as shown in Fig. 4. At the bottom of the container, a hole was made which has been connected to a transparent tube along with measuring scale. The rise of water in the container can be seen from outside in certain time. By knowing the volume of water in certain time, actual discharge can be calculated.

4 Results and Discussion

The coefficient of discharge of venturimeter has been found by experimental and computational analysis both and their results have been compared. Furthermore, effect of flow rate on coefficient of discharge of the venturimeter was also studied.

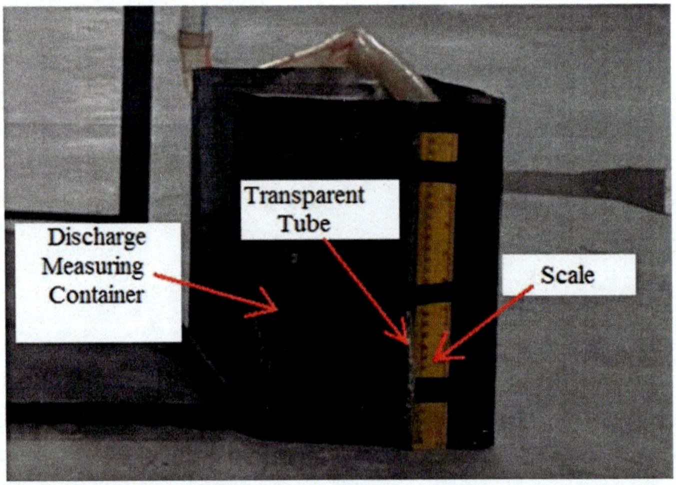

Fig. 4 Discharge measuring container

4.1 Experimental Results and Discussion

The inlet tank, i.e., sump has been filled with normal water at the room temperature before starting the experiments. The pump is started and flow rate through the venturimeter is controlled by ball type valve. After setting the certain mass flow rate by ball type valve, deflection of mercury in U-tube, rise of water level in discharge measuring container and time of water rise in the measuring container was noted. This process has been repeated with incremental increase in flow rate for twelve number of tests to show the effect of discharge on Cd and all the reading was noted, which are presented in Table 1.

Theoretical discharge through the venturimeter is given by Eq. (1) as

$$Q_{\text{theoretical}} = A_1.A_2.\sqrt{\frac{2gh}{(A_1^2 - A_2^2)}}$$

where h = head difference of water level between Section-1 and Section-2, A_1 = area of pipe at Section-1, A_2 = throat area, g = acceleration due to gravity. From the Fig. 1, dimension of the venturimeter is D_1 = 27 mm = 0.027 m, D_2 = 9 mm = 0.009 m.

$$A_1 = \frac{\pi D_1^2}{4} = 5.73 \times 10^{-4}\text{m}^2, \quad A_2 = \frac{\pi D_2^2}{4} = 6.36 \times 10^{-5}\text{m}^2$$

Hence, theoretical discharge in the term of h is given by

$$Q_{\text{theoretical}} = 28.36967 \times 10^{-5}\sqrt{h(m)}$$

From the Eq. (2), we have

$$h = x.\left[\frac{\rho_{hg}}{\rho} - 1\right]$$

where x = deflection of mercury in U-tube.

Put density of mercury, ρ_{hg} = 13.6 × 1000 kg/m^3 and density of water, ρ = 1000 kg/m^3.

$$h = 12.6x.$$

Hence, theoretical discharge in the term of deflection of mercury (x) is given by

$$Q_{\text{theoretical}} = 100.7 \times 10^{-5}\sqrt{x(m)},\ \text{m}^3/\text{s}$$

Table 1 Observations and results

Test no	Deflection of mercury x (cm)	$Q_{\text{theoretical}}(\text{m}^3/\text{s}) = 10.07 \times 10^{-5}\sqrt{x(\text{cm})}$	Rise of water in tank h_{tank} (cm)	V_{tank} (m^3) = 6.25 × 10^{-4} *h_{tank} (cm)	Time of measurement, t (s)	$Q_{\text{actual}} = \frac{V_{\text{tank}}}{t}$ (m^3/s)	$C_d = \frac{Q_{\text{actual}}}{Q_{\text{theoretical}}}$
1	3.2	0.000180	20.0	0.0125	72	0.000174	0.96603
2	4.0	0.000201	20.0	0.0125	64	0.000195	0.96812
3	5.3	0.000231	20.0	0.0125	56	0.000224	0.96954
4	6.2	0.000250	20.0	0.0125	51	0.000243	0.97102
5	7.1	0.000268	20.0	0.0125	48	0.000261	0.97324
6	8.0	0.000284	20.0	0.0125	45	0.000277	0.97533
7	9.4	0.000308	20.0	0.0125	42	0.000301	0.97721
8	10.0	0.000318	20.0	0.0125	40	0.000312	0.97982
9	11.2	0.000337	20.0	0.0125	38	0.000331	0.98241
10	12.1	0.000350	20.0	0.0125	36	0.000344	0.98413
11	13.2	0.000365	20.0	0.0125	35	0.000360	0.98623
12	14.0	0.000376	20.0	0.0125	34	0.000372	0.98934

or

$$Q_{\text{theoretical}} = 10.07 \times 10^{-5}\sqrt{x(\text{cm})},\ \text{m}^3/\text{s}.$$

Volume of water collected in tank in time t, for rising of water level in tank (h_{tank}) is given by

$$V_{\text{tank}} = \text{cross sectional area of tank} \times \text{Rise of water level in the tank } (h_{\text{tank}})$$

Cross sectional area of tank = 0.25 × 0.25 m = 6.25×10^{-2} m^2.

$$V_{\text{tank}} = 6.25 \times 10^{-2} \times h_{\text{tank}}(\text{m}),\ \text{m}^3 \quad \text{or} \quad V_{\text{tank}} = 6.25 \times 10^{-4} \times h_{\text{tank}}(\text{cm}),\ \text{m}^3$$

Hence, actual discharge in m^3/s is given by,

$$Q_{\text{actual}} = \frac{V_{\text{tank}}}{t}$$

Figure 5 shows the variation of coefficient of discharge (Cd) with the discharge. From this result, it was observed that coefficient of discharge (Cd) increases with increase in discharge at the inlet. For the low discharge, Cd value is very small and for high discharge Cd value is very high. Figure 6 shows the comparative values of theoretical and actual discharge for 12 number of tests. From this result, it was found that the theoretical discharge is more than the actual discharge.

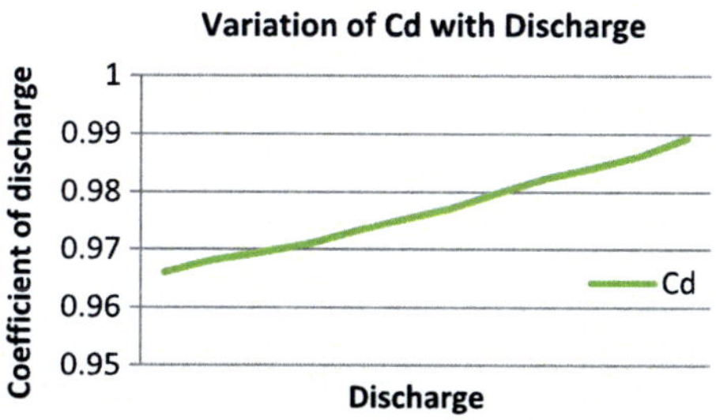

Fig. 5 Variation of C_d with discharge

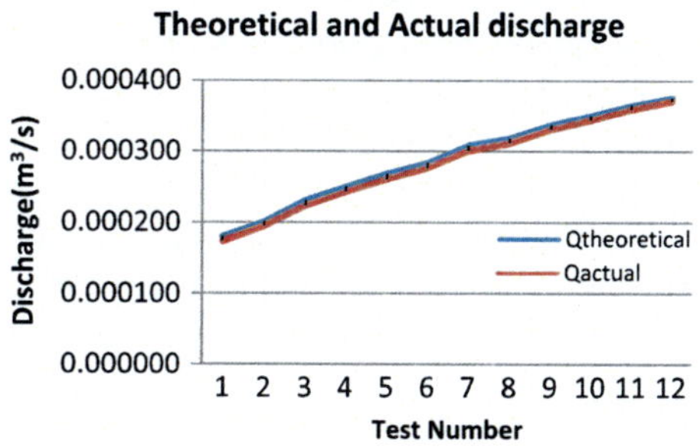

Fig. 6 Theoretical and actual discharge

Table 2 Parameters for numerical simulation

S.n	Parameter	Value
1.	Atmospheric pressure	101,320 Pa
2.	Atmospheric temperature	300 K
3.	Gravity	9.81 m/s^2
4.	Wall condition	Adiabatic wall condition
5.	Type of flow	Laminar and turbulent flow

4.2 Computational Results and Discussion

Drafting and modeling are very essential for the computational analysis of flow through venturimeter. SolidWorks 2016 was used for modeling and drafting the venturimeter. The venturimeter 2-D sketch was made in SolidWorks and extruded into 3-D model. SolidWorks flow simulation has been used for flow analysis through venturimeter. Tetrahedral face meshing was used to produce the fine mesh. For computational analysis, the parameters adopted for inlet condition of the venturimeter is presented in Table 2. The results converged in 200 numbers of iterations.

Experiments were conducted for 12 number of tests, and also, computational analysis has been done for four number of tests to compare the results. Computational results for the 4 number of tests, i.e., for 4 number of inlet condition are given below.

Test Number-4, Velocity and Pressure Contour

For inlet condition, velocity $V_1 = 0.4363$ m/s, pressure $P_1 = 113.79$ kPa, $Q_1 = 0.000250$ m^3/s is shown below.

Test Number-8, Velocity and Pressure Contour

For inlet condition, velocity $V_2 = 0.5549$ m/s, pressure $P_2 = 127.58$ kPa, $Q_2 = 0.000318$ m^3/s is shown below.

Test Number-10, Velocity and Pressure Contour

For inlet condition, velocity $V_3 = 0.6108$ m/s, pressure $P_3 = 141$ kPa, $Q_3 = 0.000350$ m^3/s is shown below.

Test Number-12, Velocity and Pressure Contour

For inlet condition, velocity $V_4 = 0.6561$ m/s, pressure $P_4 = 155.17$ kPa, $Q_4 = 0.000376$ m^3/s are shown below.

From Figs. 7, 8, 9, 10, 11, 12, 13, 14, velocity and pressure contour has been presented for four different inlet conditions, as discussed above, for the selected venturimeter.

For the test number-4, from the velocity contour, the average velocity is found to be 3.923 m/s. Hence, computational discharge can be calculated as

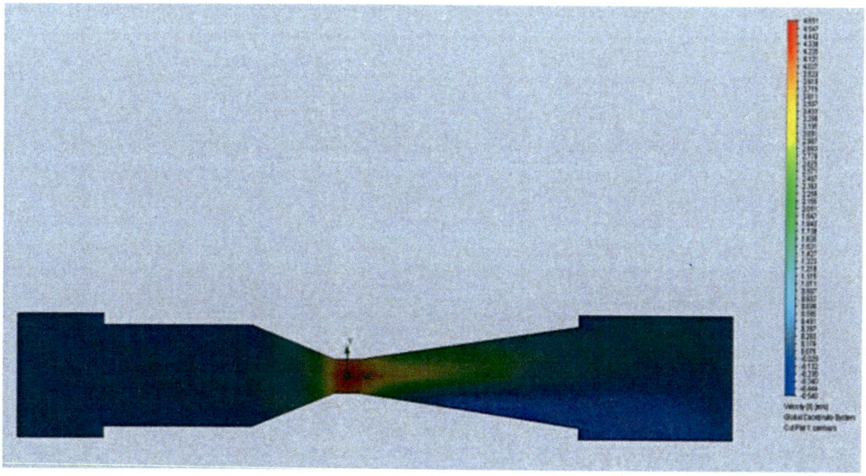

Fig. 7 Velocity contour for test-4

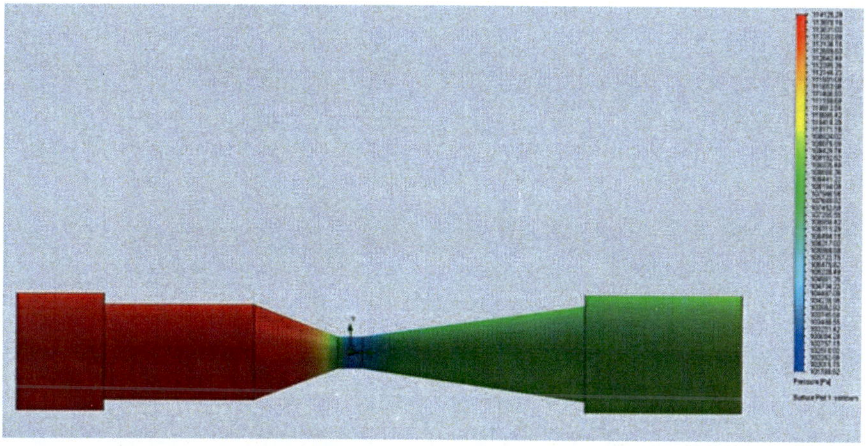

Fig. 8 Pressure contour for test-4

$$Q_{\text{comp}} = A_2.V_2 = 6.36 \times 10^{-5} \times 3.923 = 0.000249\ \text{m}^3/\text{s},\quad Q_{\text{actual}} = 0.000243$$

The coefficient of discharge, from the computational analysis is

$$(C_d)_{\text{comp}} = Q_{\text{actual}}/\ Q_{\text{comp}} = 0.000243/0.000249 = 0.9759$$

For the test number-12, from the velocity contour, the average velocity is found to be 5.89 m/s. Hence, the computational discharge can be found as

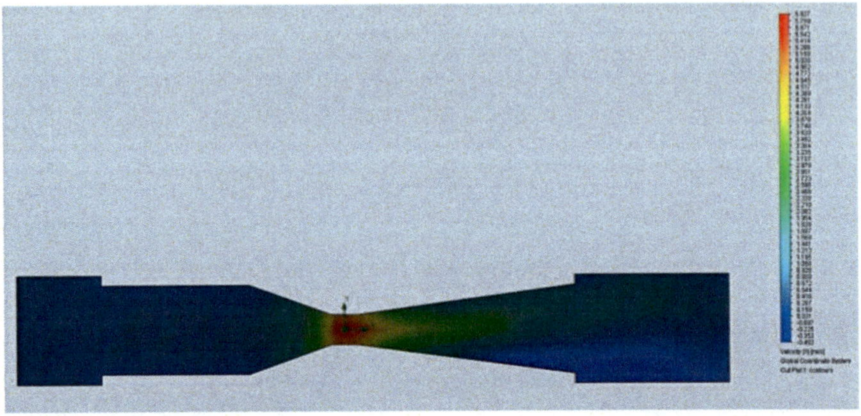

Fig. 9 Velocity contour for test-8

Fig. 10 Pressure contour for test-8

$$Q_{\text{comp}} = A_2.V_2 = 6.36 \times 10^{-5} \times 5.89 = 0.000375\ \text{m}^3/\text{s}, \quad Q_{\text{actual}} = 0.000372$$

The coefficient of discharge, from the computational analysis, is

$$(C_d)_{\text{comp}} = Q_{\text{actual}} / Q_{\text{comp}} = 0.000372/0.000375 = 0.9920$$

For the test number-4, i.e., for low discharge, (Table 1), the experimental value of coefficient of discharge is found to be 0.9710, and from the computational analysis, the coefficient of discharge is found to be 0.9759. Furthermore, for the test number-12, i.e., for high discharge, the experimental value of coefficient of discharge is found to be 0.9893. From the computational analysis, the coefficient of discharge is found to be 0.9920. Hence, percentage change in coefficient of discharge obtained from the numerical analysis and experimental results for low and

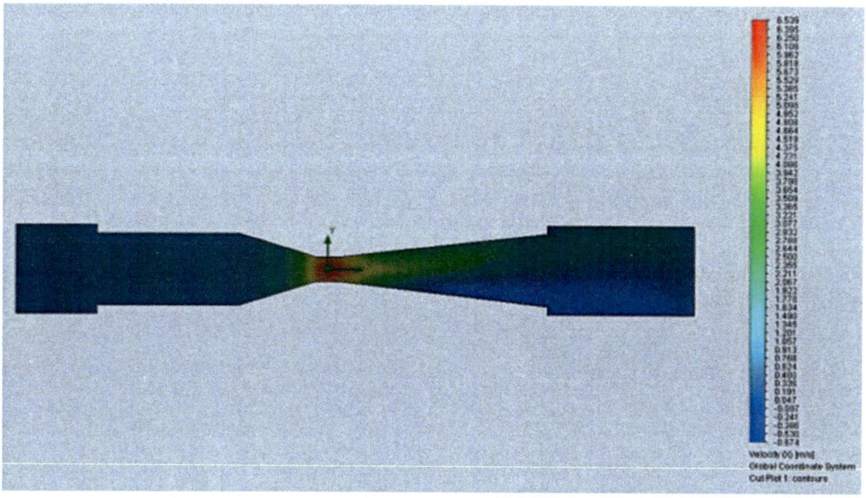

Fig. 11 Velocity contour for test-10

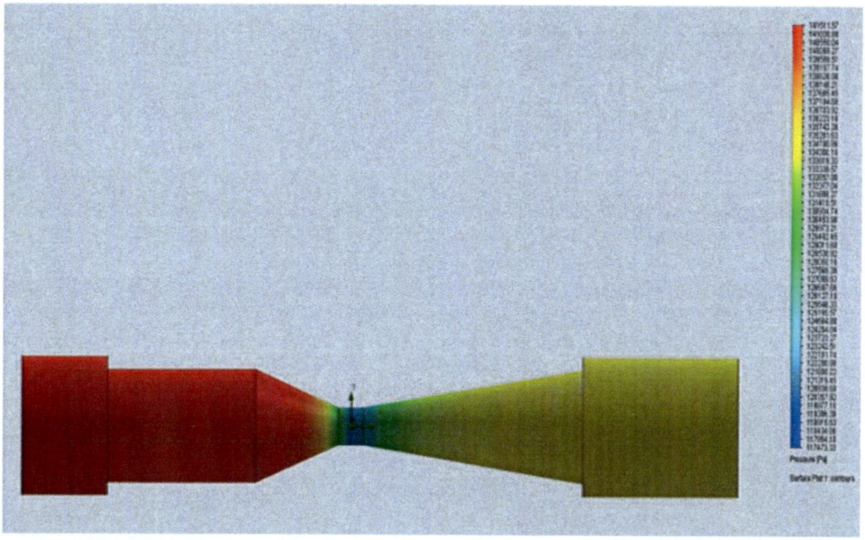

Fig. 12 Pressure contour for test-10

high discharge, respectively, was found to be 0.50% and 0.28% which is less than 1%. This difference may be due to friction, surface roughness and sharp edges at the joint.

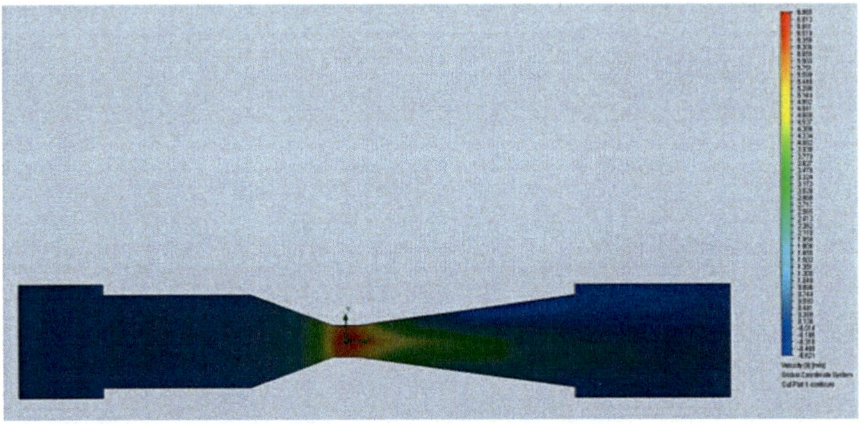

Fig. 13 Velocity contour for test-12

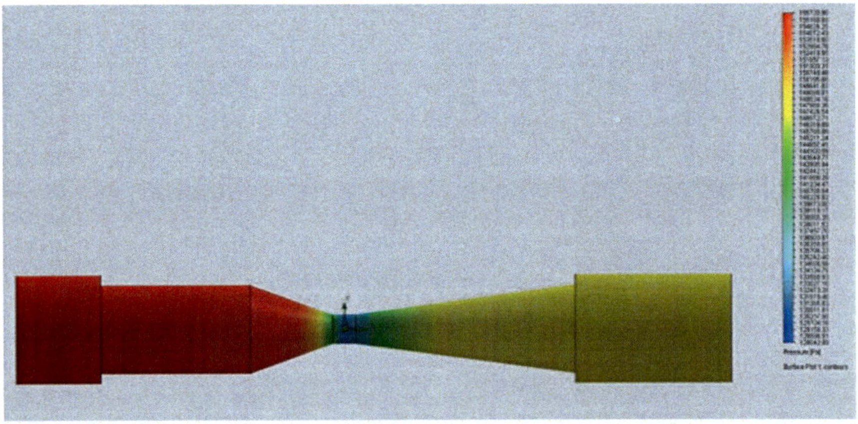

Fig. 14 Pressure contour for test-12

5 Conclusions

In the present work, experimental and computational analysis of coefficient of discharge (Cd) of venturimeter is studied extensively. For computational analysis, venturimeter is designed and simulated using SolidWorks software and for experimental measurements, the venturimeter is fabricated as per proposed design. The effect of flow rate on the coefficient of discharge was studied using experimental analysis, and the results were validated computationally using SolidWorks 2016 software. The following conclusions are made:

1. Experimental results revealed that the coefficient of discharge (C_d) increases with increase in rate of flow, i.e., increase in Reynolds number.
2. For high flow rate, the coefficient of discharge was found as 0.9893 and computational value of C_d was found as 0.992.
3. The coefficient of discharge computed experimentally and computationally, varied by less than one percent. This indicates that the design parameters of the venturimeter were correctly chosen to get accurate results.

The results presented in this paper provide a baseline to design the venturimeter for measuring the flow rate in the devices where high flow rates are expected. Furthermore, the optimum range of divergence angle of the diverging section of the venturimeter can be set between 4° and 7°.

References

1. Bansal, R.K.: A Text Book of Fluid Mechanics & Hydraulics Machines. Laxmi Publications Pvt. Ltd. (2010) (revised)
2. Daugherty, R.L.: Fluid Mechanics with Engineering Applications. Tata McGraw-Hill Education, New York (1989)
3. Nithin, T., Jain, N., Hiriyannaiah, A.: Optimization of venturi flow meter model for the angle of divergence with minimal pressure drop by computational fluid dynamics method. In: International Conference on Challenges and Opportunities in Mechanical Engineering, Industrial Engineering and Management Studies, July 2012
4. Suryawanshi, M., Bhinge, A. Kokane, A., Dhaigude, S., Bajaj, R.: Effects of geometrical parameters on performance characteristics of venturimeter. Int. J. Theor. Appl. Res. Mech. Eng. **6**(1–2), 2319–3182 (2017)
5. Sanghani, C., Jayani, D.C., Jadvani, N.R., Dobariya, H., Jasoliya, K.R.: Effect of geometrical parameters of venturimeter on pressure drop. Int. J. Sci. Res. Sci. Eng. Technol. **2**, 865–868 (2016)
6. Kumar, J., Singh, J., Harsh, K., Narula, G.S., Singh, P.: CFD analysis of flow through venturi. Int. J. Res. Mech. Eng. Technol. **4**, 214–217 (2014)
7. Arun, R., Yogesh Kumar, K.J., Seshadri, V.: Prediction of discharge coefficient of venturimeter at low Reynolds numbers by analytical and CFD method. Int. J. Eng. Tech. Res (IJETR). **3**(5) (2015). ISSN: 2321–0869
8. Zhang, J.X.: Analysis on the effect of venturi tube structural parameters on fluid flow. AIP Adv. **7**(6) (2017)
9. Yadav, O., Gaikwad, P.: Swirling flow visualization in a square section test duct by particle image velocimetry (PIV) system. Int. J. Adv. Mech. Eng. **4**(2), 175–184 (2014)
10. Tukimin, A., Zuber, M., Ahmad, K.A.: CFD analysis of flow through venturi tube and its discharge coefficient. In: IOP Conference Series: Materials Science and Engineering, vol. 152, no. 1, p. 012062. IOP Publishing (2016)
11. Gupta, B., Nayak, A.K., Kandar, T.K., Nair, S.: Investigation of air-water two phase flow through a venturi. Exp. Therm. Fluid. Sci. **70**, 148–154 (2016)
12. Akpan, P.: A CFD simulation of water flow through a venturi meter. Int. J. Curr. Res. **6**, 5425–5431 (2014)
13. Chung, T.J.: Computational Fluid Dynamics, 2nd edn (2010). https://doi.org/10.1017/CBO9780511780066

14. Tamhankar, N., Pandhare, A., Joglekar, A., Bansode, V.: Experimental and CFD analysis of flow through venturimeter to determine the coefficient of discharge. Int. J. Latest Trends Eng. Technol. **3**(4), 194–200 (2014)
15. Yadav, O., Yadav, V.K., Tiwari, S.: Experimental investigation of counter flow heat exchanger with swirling. Int. J. Eng. Technol. **7**(4.39), 169–171 (2018)
16. Yadav, V.K., Ranjan, P., Yadav, V., Yadav, O.: Flame structures of burner stabilized laminar premixed flames. Int. J. Eng. Technol. **7**(4.39), 172–176 (2018)

A Comprehensive Review of Performance, Combustion, and Emission Characteristics of Biodiesel Blend with Nanoparticles in Diesel Engines

Suraj Bhan, Raghvendra Gautam, Pusphendra Singh, and Abhishek Sharma

Nomenclature

ASTM	American standard test method
BTE	Brake thermal efficiency
CO_2	Carbon Dioxide
BP	Brake Power
B5KOEE95	Diesel 95% + Kusum oil ethyl ester 5%
D100	Diesel 100%
JOME5	Jatropha oil methyl ester 5% + 95% diesel
JOEE5	Jatropha oil ethyl ester 10% + 90% diesel
ASTM	American Standard Test Method
η (Mech)	Mechanical Efficiency
CRDI	Common Rail Direct Injection
TDC	Before Top Dead Centre
CN	Cetane Number
PM	Particulate Matters
RCCD	Rotable central composite design
SME	Sunflower methyl ester
KOH	Potassium Hydroxide
CI	Compression Ignition
NO_X	Oxides of Nitrogen
CO	Carbon mono oxide
B10KOEE90	Diesel 90% + Kusum oil ethyl ester
B15KOEE95	Diesel 85% + Kusum oil ethyl ester 15%
B20KOEE80	Diesel 80% + Kusum oil ethyl ester 20%

S. Bhan (✉) · R. Gautam · P. Singh
Department of Mechanical Engineering, Delhi Technological University, Delhi, India

A. Sharma
Department of Mechanical Engineering, G L Bajaj Institute of Technology and Management, Greater Noida, UP, India

L. M. Das et al. (eds.), *Recent Trends in Thermal Engineering*,
Lecture Notes in Mechanical Engineering,
https://doi.org/10.1007/978-981-16-3428-4_7

JOME10	Jatropha oil methyl ester 10% + 90% diesel
JOEE5	Jatropha oil ethyl ester 5% + 95% diesel
F/A ratio	Fuel Air ratio
CME	Canola methyl ester
NaOH	Sodium Hydroxide
CV	Calorific Value
BEMP	Brake Effective Mean Pressure
RME	Rapeseed methyl ester
LTC	Low temperature combustion
BTE	Beef tallow ester
DAG	Diacylglycerols
FAME	Free acid methyl ester
FIP	Fuel Injection Pressure

1 Introduction

The energy request is expanding step-by-step and oil subsidiaries are contributing a considerable amount of its expansion especially in the transportation division. The quest for a substitute fuel has provoked various revelations, because of that a wide variety of elective energizes are accessible for transportation, force, energy, and rural purposes [1, 2]. Diesel engines are widely preferred to utilize fuel energy because of its high-thermal efficiency comparatively. One of the principal issues related to diesel engines is its harmful discharges, for example, carbon dioxide (CO_2), carbon monoxide (CO), particulate issue (PM), soot, and unburned hydrocarbons (HC). These discharges pollute the atmosphere which is the basic problem associated with human beings living. Particulate issues in the diesel engines fumes are exceptionally risky as it influences the lungs and has potential impacts on heart, though carbon monoxide joins with the hemoglobin in the blood, confines the progression of oxygen subsequently prompting unexpected passing [3]. HC and NO_X discharge within the sight of daylight respond to shape photochemical smog and acid rain which causes respiratory issues and eye bothering. The particulate issue makes bothering the eyes, nose, throat, and lungs, bringing about cardiovascular diseases and potentially sudden passing [3].

With the fast exhaustion of petroleum product concern is developing on its proceeded with accessibility and about its harmful side of emissions due to which our planet polluted for humankind and henceforth adequately draw the attention about an alternate with a substitute wellspring of energy which out and out quickened the exploration on biodiesel [4]. Replacement of petroleum derivative with fuel extracted through various types of plants seeds and fats of animals and algae may be a better choice as the biodiesel has a great element, for example, fair on execution, favorable emissions, tremendous accessibility, locally accessible and its sustainable nature has additionally tuned the expanding center around biodiesel [5]. Biodiesel has overpowering characters like biodegradability, better lubricity, eco-cordiality, and non-harmfulness when utilized in diesel engines. Accordingly, biodiesel may be a possible expected choice for diesel. Second era research paid attention just too non-consumable oil in light of food necessity cost worry. In the investigation of substitute fuel challenges around the world, numerous experiments were performed with biofuel. It was found that biofuel may be alternative fuel [6].

Many researchers are serious to compensate for the energy crisis and reduce the harmful emissions to produce a healthy atmosphere. So they are in search of alternative fuel which may lessen the utilization of diesel furthermore for the decrease of hurtful emissions. Biofuel may be proving the better alternative fuel. Biofuel can be produced from a variety of vegetable oils, fats of the creature, and algae. Vegetable oil can be obtained from rapeseed, soybean, palm; sunflower, etc. researchers are using a wide range of technology to reduce emissions as catalysts of diesel oxidation, particulate matter filters (PMF), and exhaust gas recirculation (EGR). Many researchers study that biofuel blending with diesel reduces emissions like CO, CO_2, HC, particulate matter, but usually increases the NO_X emissions. To reduce the NO_X, new combustion chamber technology has been using as LTC, RCCI, etc. but using LTC, NO_X reduced, but the efficiency of the engine decreased which is a challenging part for the researcher.

NO_X emissions in a diesel engine can be decreased by EGR [7, 8]. Instead of it, many researchers used the two ways catalytic converter in the exhaust system of diesel engine diminishes hydrocarbon and carbon monoxide emissions.

Still, NO_X emissions reduction is challenging for the researchers because LTC reduced the NO_X emissions but on the other aspect, it decreased the BTE, Brake power. EGR is not much efficient. Instead of it, researchers are in favor of a blend of fuel such that changes in characteristics of fuel favorable in the reduction of CO, CO_2, HC, and NO_X emissions. Biofuel blending usually increases the NO_X, but reduces emissions like CO, CO_2, and HC, etc. [9, 10].

Methanol may be used as an alternative biofuel. Methanol has high-oxygen content. It has higher oxygen and octane number compare to ethanol. These properties help to produce better combustion characteristics. Researchers find the performance and emission characteristics of methanol as a blended fuel. BTE decreases with an increase in the concentration of methanol and in another aspect BSFC increases with the increase in the concentration of methanol. The emission characteristics like CO emissions increase in comparison to the pure gasoline, but NO_X and UHC emission characteristics decrease in the case of methanol blend because of combustion of fuel occurred at a lower temperature [11].

It was seen by the researcher that BTE decreases at higher peak load by increases the blend value of neem oil by more than 5% in diesel. This may be due to the higher viscosity of fuel due to which improper atomization of fuel takes place inside the combustion chamber. But it was also observed that DN 10 and DN 15 have the higher BTE in comparison to the DN5 and DN 100 at part loads up to 60% load. Airflow remains constant at higher load, but fuel injection rate increases. There is very little time for fuel combustion therefore CO_2 and UHC increased. BSFC increases with increases lend of neem oil. This may be due to the higher density and lower calorific value of the fuel [12].

Brake thermal efficiency is defined as the ratio of brake power produced total energy of fuel. I fort was seen by the researcher that BTE improve y increasing the load up to 60%, but for pure Kusum oil, BTE decreases due to higher viscosity resulting in improper atomization of fuel.io BTE improved by mixing of butanol. Butanol is an oxygen-rich fuel and improved viscosity due to which proper combustion of fuel takes place inside the combustion chamber. BSFC increases with increases the load because it has a higher density and lower calorific value of the fuel. A lower calorific value requires a more amount of fuel for the same power production [13].

Biodiesel has higher viscosity due to which it cannot directly use in a diesel engine. The transesterification process is used to reduce the viscosity of biofuel. Due to higher viscosity, brake specific consumption, and availability of biofuel, it is used with diesel. Nanoparticles may be used as a catalyst in biofuel because of it has numerous benefits like superior density, improved combustion by increasing evaporation rate of fuel and air mixing resulting in reduced ignition delay, decreased emission like CO, CO_2, HC, and NO_X. Novel nano added substance is eminent as a superior fuel catalyst material in these ongoing hundreds of years. Oxidation and evaporation rate of biofuel increases when nanoparticle mixes in biofuel because of nanoparticle has the bigger surface area to volume ratio [14]. It was seen by the researcher addition of biofuel improves the properties of biofuel like an advanced flash point, kinematic viscosity. These properties improve due to the high surface to volume nature of nano particles [15].

To improve the engine performance and engine characteristics nanoparticles are mixed in biodiesel as an additive. The nanoparticle has a high surface to volume ratio due to which nanoparticle has the several advantages over without nanoparticle biodiesel like as improves the atomization of the fuel, allows more fuel to react with oxygen, reduces ignition delay, enhances rapid evaporation, provide better air-fuel mixing, reduced cylinder pressure, and reduced heat release rate. It was seen by the researchers that the addition of nanoparticles in biodiesel improved the brake thermal efficiency, reduced BSFC, reduced NO_X and improved other emission characteristics like as the HC, CO, and CO_2 due to following reasons:

- Nanopartcile addition in biofuel improves physio-chemical properties of fuel which is shown in Table 1.
- Nanoparticles addition reduces viscosity of blend fuel which leads to fine atomization, dispersion and mixing of fuel inside combustion chamber. So that proper combustion takes place inside combustion chamber.
- Nanoparticle has high surface to volume ratio due to which they provide more contact surface for chemical reactivity which in turn increases evaporation rate thereby increasing heat transfer. Nanoparticle improves the thermo-chemical properties of combustion of air-fuel mixture.
- Nanoparticle addition in biofuel increases calorific value of blend fuel ultimately increases the BTE of compression ignition engine.

Table 1 Change in physio-chemical properties after addition of nanoparticle

S.N.	Fuel	Composition of additives	Flash point	Kinematic viscosity	Density kg/m^3	Calorific value (MJ/Kg)	Cetane number	Reference
1	Neat diesel		48	2.20	835	42.30	–	[21]
2	Neat biofuel		85	4.10	873	39.40	–	
3	JBD30A	30 ppm Al_2O_3	78	4.25	875	38.90	–	
4	JBD30C	30 ppm CeO_2	76	4.30	876	38.70	–	
5	Diesel		56	3	815	42	47	[22]
6	Mahua methyl ester (MME)		136	4.90	869	40	56	
7	MME20		76	3.40	826	41.62	49	
8	MME20 + ANP50	ANP 50 ppm	71	3.37	827.5	41.66	50	
9	MME20 + ANP100	100 ppm	65	3.33	829	41.70	51	
10	Mahua biodiesel		144	4.5	0.98 150c (g/cc)	39.22	55	[23]
11	MB100CuO10	100 ppm 10 nm	144	4.2	0.92	39.7	56	
12	MB100CuO20	100 ppm 20 nm	143	3.7	0.88	40.128	58	
13	Neat diesel		48	2.20	835	42.30		[21]
14	Neat biodiesel		85	4.10	873	39.5		
15	JBD30A		78	4.25	875	38.9		
16	JBD30C		76	4.30	876	38.7		
17	Palm oil methyl ester		196	4.2	0.877 g/ml			[24]
18	POM2O		71.5	2.82	0.835	41.5		
19	POM20 + 30PPM	CeO_2	52	3.58 CST	862	42		

- Nanopartcile accelerated the evaporation rate of air-fuel mixture and reduces the pre mixed fraction of air-fuel mixture due to which ignition delay and combustion temperature reduced. The presence of lower active radicals led to a decrease in the potential of thermal NO_X formation.
- Nanopartcile improve cetane value, calorifc value, reduced viscosity and improve combustion of fuel due to which BSFC of biofuel reduced [16–20].

It was seen by researcher that nanoparticle blend in biofuel form agglomeration of nanoparticles due to minute size of nanoparticle. Nanoparticle may create sedimentation problem. To maintain the stability and homogenous mixture ultrasonicator are sed at high frequency. Surfactants (Cetyltrimethyl ammonium bromide) may be used to control the sedimentation problem and avoid the agglomeration of nanoparticles. To characterize the nanoparticle various technique are used such as

- **X-ray diffraction (XRD)**: used to identify the crystalline phase and to determine the crystalline size of nanoparticles
- **Scanning electron microscopy (SEM)**: used to measure the morphology and average particle size of nanoparticles
- **Energy dispersive spectrum (EDS)**: used to check of the composition of nanoparticles [16–20].

Recently, nanoparticles are used as an additive with biofuel. Nanoparticles improve the characteristics of the biodiesel and get better the diesel engine performance. The nanoparticle may increase the thermal conductivity, diffusivity, calorific value, density of the fuel resulting in better combustion take place inside the combustion chamber. Nanoparticle reduces the ignition delay because nanoparticle has a high surface to volume ratio which means more contact surface area between fuel and oxidizer. Due to which nanoparticle increases the evaporation rate between fuel-air mixing resulting accelerates deeply the combustion flame and early combustion occurred comparatively. So as this type of ignition delay period was reduced. On the other aspect, it increases the density of nanoparticle due to which it increases the injection velocity of the fuel resolution to improve the atomization of fuel. Nanoparticle addition with biodiesel improves the combustion resulting in reducing emissions like CO, CO_2, NO_X, HC, and get better the overall engine performance as BTE, BSFC, etc. [25].

It was seen by the researcher that 5–100 nm particle size is suitable as an additive with biodiesel. Nanoparticles are formed by various chemical and physical processes. Many metal and metal oxides are usually used as the nanoparticle ex Aluminum, Titanium, Copper, Cerium, Boron, Iron, Zinc, etc., and their oxides. There is much experimental work is carried out to improve the characteristics of biodiesel and the emission and performance of the engine. In this review paper, more experimental studies have been summarizing, and investigate gaps where additional exploration is desirable.

There are many methods to optimize the different variables in biodiesel blending. The researcher used Taguchi approach to optimize the reaction variables for the reduction of FFA. Temperature, catalyst amount, molar ratio, and reaction time variables studies to optimize the FFA because these variables affect the acid value during the transesterification process. Researcher finds that maximum yield obtained at molar ratio 5:1, 1% catalyst, 65 °C temperature of reaction, and 120 min reaction time by the average S/N ratio. At these parameters highest yielding obtained and the lowest reduction of FFA [26].

2 Effect of Nanoparticles on Performance and Emission Characteristics of a Diesel Engine

To improve the engine performance and engine characteristics nanoparticles are mixed in biodiesel as an additive. The nanoparticle has a high surface to volume ratio due to which nanoparticle has the several advantages over without nanoparticle biodiesel like as improves the atomization of the fuel, allows more fuel to react with oxygen, reduces ignition delay, enhances rapid evaporation, provide better air-fuel mixing, reduced cylinder pressure, and reduced heat release rate.

It was seen by the researchers that the addition of nanoparticles in biodiesel improved the brake thermal efficiency in comparison to 100% biofuel. It was analyzed experimentally that there are more improvements in BTE for B20A30C30 and B100A30C30 test fuel in comparison to pure 100% biofuel. Advance combustion takes place in the combustion chamber when nanoparticles are used as an additive in biodiesel because the nanoparticle short the ignition delay. Nanoparticles have a high surface and volume ratio due to which the evaporation rate of fuel increased. Due to advanced combustion, it was seen that there are low-combustion chamber pressure and heat release rate at full load conditions. The NO discharge radically diminished for the B20A30C30 test fuel with a rate decrease in 30%, when compared with B100 because of the consolidated impact of Alumina and Cerium oxide nanoparticles. It was seen by the researcher that CO, UBHC, and smoke opacity are reduced due to an improved oxidation-reduction process when nanoparticles are mixed in biodiesel [27].

Magnalium and Cobalt Oxide are used as nanoparticles in Jatropha biodiesel fuel to check diesel engine performance. It was found that the content of nanoparticles brought about a compelling upgrading in the BTE and decrement in the BSFC by 2%. It was seen that the emanations were diminished viably as UHC was decreased by 60%, CO by half, and the NO_X by 45% [28] (Fig. 1).

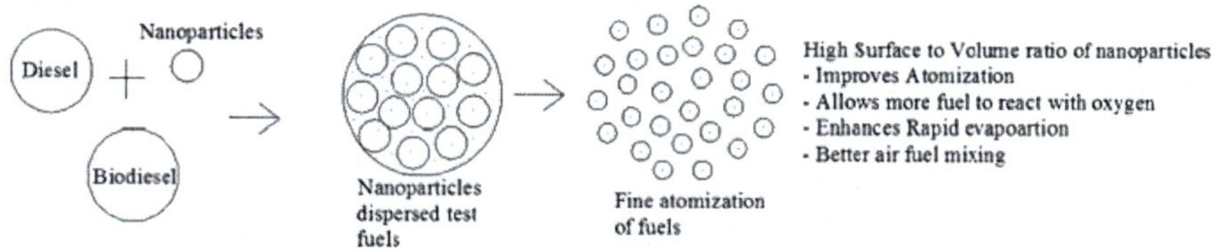

Fig. 1 Dispersion effect of nanoparticle with biofuel blending [27]

Researchers researched the effects of Cerium oxide used as an additive substance in diesel furthermore; diesel-biodiesel-ethanol mixes to check the diesel engine performance. It was seen by the researcher that there is a reduction in specific fuel consumption and ignition delay after the addition of nanoparticles in biofuel. The adding up of Cerium oxide likewise quickened before the commencement of burning and caused an inferior heat discharge rate. It was seen that CO, UHC, and NO_X emissions effectively decreased. Researcher likewise revealed that the best possible engine performance was accomplished at nanoparticles portion level of 35 ppm [29–31].

Researchers considered the effect of adding nano- and micro-sized aluminum (Al) particle size, dispersant focus, and base fluid type to change its effect on the combustion characteristics of n- decane and ethanol droplets. They found that the engine power has been amazingly improved. They also detailed the effective reduction of carbon dioxide and nitrogen dioxide [32].

The researcher observed the emission characteristics and engine performance of diesel engines when ferric chloride ($FeCl_3$) was used as a nanoparticle additive in biodiesel. Due to the effect of nanoparticle blending BSFC decreased to 8.6% and brake thermal efficiency increased to 6.30%. The researcher also found that cylinder means effective pressure and hear radiate rate increased effectively and emission characteristics like as NO_X, CO, UHC, and smoke were reduced by 21.60% 52.6%, 27.0%, and 7.0%, respectively [33].

Aluminum oxide (Al_2O_3) and Copper oxide (CuO) used as nanoparticles additive to check the performance of diesel engine and emission characteristics. Flashpoint and cetane number increased to 1% and 3.3% when Aluminum oxide (Al_2O_3) and Copper oxide (CuO) nanoparticles were mixed with biodiesel. It was seen by the researcher that engine emissions like NO_X, CO, and UHC were fundamentally decreased by 11%, 13%, and 6%, separately [34].

When jajoba biofuel blended with the diesel the peak pressure decreased in comparison to diesel fuel at 75% load at various speeds. This is due to the superior viscosity of the biofuel which leads to improper atomization of the fuel resulting in improper combustion take place inside the combustion chamber. It was also seen that retardation in the peak value of pressure attributed to an increase in ignition delay. Due to an increase in ignition delay, more amount of fuel accumulated for burning which leads to improper combustion. By mixing of nanoparticle as Al_2O_3 into the JB20D mixture evaporation rate of the fuel droplet increases which improve the combustion process because of Nanoparticles (Al_2O_3) has the higher surface to volume ratio and higher thermal conductivity. Nanoparticle mixing reduces the ignition delay because of combustion of fuel occurred early comparatively without nano fuel blending. It was seen that peak pressure also increased inside the combustion chamber by mixing the nanoparticles [35].

The rate of pressure rise curve indicates the rate of heat released and the degree of the combustion reaction. So to obtain the maximum rate of pressure rise of the curve is an important research design parameter of the engine [36]. It was seen by the researcher adding the nanoparticles peak pressure rise rate increases due to better fuel mixing resulting in better evaporation rate of fuel.

Ignition delay is a very important parameter that affects the performance of the engine. Biofuel has an increased viscosity in comparison to diesel fuel. There are some limitations due to the high viscosity of biofuel like fuel spray in larger droplets form which leads to improper combustion inside the cylinder. On another hand, carbon deposits formed in the nozzle of the injector and cylinder liner due to the higher viscosity of the nozzle [37, 38].

At lower injection pressures denser or say larger fuel droplets do not mix uniformly with compressed air so that uneven combustion takes place inside the combustion chamber. Due to which evaporation rate affected and ignition delay period increases resulting in more fuel accumulated inside the combustion chamber which leads to knocking. To avoid knocking, shorter ignition delay is required.

If injection pressure increased, fuel spray, dispersion, and mixing of fuel with air properly take place because fuel split into very small droplets. Fuel spray travels more distance comparatively when injection pressure of fuel in nozzle increased. There is more distance covered by the fuel droplet by increasing injection pressure parameters. When fuel droplets covered more distance, fuel droplets properly mix with the high pressure and temperature compressed air. Properly mixing increases the evaporation rate of fuel ultimately reduce the ignition delay time and CO emissions and proper combustion take place inside the combustion chamber and give a higher heat release rate and reduced emissions except for NO_X. But it was seen that to increase the injection pressure the BTE increased usually but BSFC increased. So to decrease the oxides of nitrogen and BSFC nano particles may be mixed with the biofuel with higher injection pressure [39].

Biofuel has a higher viscosity due to which proper atomization, penetration, dispersion, and thermal mixing of air-fuel affected. These factors are sufficient to increase the delay period. So the combustion of fluid is initiated later and maybe continuous in expansion stroke which can reduce the work output of the engine. If the injection pressure is increased, proper atomization of fuel takes place inside the combustion chamber. The density of fuel decreased because of larger fuel droplet breaks into smaller fuel droplets. Lower density fuel can easily penetrate and properly thermal mixed with air and improve the combustion. But it was seen by the researcher that BSFC decreased due to high injection pressure and NO_X increased due to more oxygen molecules into the biofuel structure. To improve combustion, reduce ignition delay, reduce BSFC, and reduce NO_X nanoparticle CeO_2 mixed with the biofuel at higher injection pressure. CeO_2 has a higher thermal conductivity die to which that enhances the mean combustion pressure and heat transfer rate between air and fuel. Therefore, this type of blending of nanoparticle in fuel increases fuel evaporation and prepares fuel in advance for auto –ignition, thereby reducing ignition delay. It was observed that for 240 bar IP with 80 ppm nanoparticles, the delay time reduced in comparison to pure B 20 fuel without using nanoparticles.

Nanoparticle CeO_2 and higher injection pressure reduce the ignition delay and improve the combustion characteristics due to which BTE increased and BSFC decreased. It was seen by many researchers that by lending of biofuel NO_X increased because its molecular structure contains more oxygen resulting in fuel burns at the higher temperature. On the other aspect nanoparticle CeO_2 mixing with biodiesel blending reduce the NO_X. Cerium oxide (Ce_2O_3) produced during the decrease of hydrocarbons might be re-oxidized to CeO_2 by the decrease of nitrogen oxide, in this manner bringing down the development of oxides of nitrogen. Cerium oxide works as an oxygen shock absorber, causing concurrent oxidation of hydrocarbons just as the decrease in nitrous oxides. The reaction can be written as follows [40]:

$$Ce_2O_3 + NO \rightarrow 2Ce_2O_2 + 1/2N_2 \quad (1)$$

Equation 1 Cerium oxide nanoparticle reaction with oxide of nitrogen.

When nanoparticle TiO_2 is mixed with the biofuel mahua oil, there is a significant reduction in the NO_X, CO, HC, and smoke opacity emissions at all braking power because nanoparticle TiO_2 has higher thermal conductivity resulting in lower ignition delay. On the other aspect, Nanoparticle has a high surface to volume ratio due to fuel properly mix with air and increase the evaporation rate and reduce ignition delay. By adding 200 ppm of TiO_2 nanoparticles to BD100, NO_X emissions can be reduced by 6.6% at peak braking power. The NO_X emissions of BD100, BD100T100, BD100T200, and diesel at peak braking power are 13.3, 13, 12.8, and 11.8 g/kWh, respectively [41].

Nanoparticles addition in biofuel improves the brake specific fuel consumption. It was seen by the researcher that when Al nanoparticles are blended in biofuel J20, there was a decrement in specific fuel consumption (6.5%) in comparison to standard diesel at 75% load conditions.

Nanoparticle blending improves fuel droplet propagation and fuel scattering inside the combustion chamber. Nanoparticle reduces the viscosity of fuel due to which fuel injected into fine particles comparatively and exposes the higher effective area. Nanoparticle provides the high surface to volume ratio resulting in proper mixing of fuel and air and works as oxygen donor resulting proper combustion takes place inside the combustion chamber. Nanoparticle increases the evaporation rate of the fuel and decreases the ignition delay and expansion of nanoparticles prompts oxidize the carbon particle properly due to which less fuel consumption takes place. Addition of nanoparticles provides the larger surface to volume ratio due to which heat transfer air-fuel mixing increased. Lesser ignition delay reduces the combustion duration of the fuel resulting in higher peak pressure and increases heat release rate. Nanoparticle addition increases the density of fuel due to which fuel injection velocity increases resulting also in fuel droplet propagation and more injected fuel dispersion. Nanoparticles contain a high surface for the chemical reaction due to which reactivity of the fuel is the main factor responsible for the reduction in ignition delay. It was seen that when nanoparticle mix with biofuel BTE increases. J20Al100 made the most extreme increment in thermal efficiency up to 6.5% at 75% of engine load [42–45].

It was found that NO_X emissions for the biodiesel mix are higher than that for the perfect diesel. The diesel engine produces more NO_X emissions for biofuel compare to the standard diesel engine at a higher engine load. Adiabatic flame temperature of combustion and cylinder temperature increased at higher engine load. At the higher temperature, more oxides of nitrogen are produced. Higher burning temperatures, accessibility of oxygen, and lower ignition delay support in producing the higher NO_X. Biodiesel contains more oxygen molecules in molecular structure due to which proper combustion occurred at higher temperature resulted in more NO_X produced. NOx and CO discharge decrease through the expansion in the more quantity of nanoparticles because of the impact of oxygenated added substances and burning upgrade. Expansion of oxides of nano-metal prompts total burning, as it goes about as an oxygen-impetus due to which CO also decreased. The most extreme heat discharge rate furthermore, high peak pressure during the ignition is normal in such a case. Premixing timing of air-fuel for burning, chamber ignition temperature, and NO_X discharge decrease by mixing of nanoparticle in biofuel because of ignition delay period decreased. The expansion of the nanoparticles to Jatropha biodiesel prompted a decline in NO_X discharges in comparison with the Jatropha biodiesel mix and diesel oil because of the catalyst impact of the nanoparticles. Nanoparticles accelerated the proper combustion and reduce the ignition delay. The greatest decrease in NO_X and CO emissions up to 52% at 75% of engine load and 35% at 75% of engine load, respectively, for J20C50 is experiential compared with all different fuel tested.

Nanoparticles get a better rate of fuel-air blending and identical burning is prompting better combustions. Nanoparticle blending mixing decreases the viscosity and gives better atomization of fuel resulting decrease in HC and co emissions. Ignition rate and combustion period improved due to the presence of more oxygen molecule in biodiesel.

Lower hydrocarbons emissions were likewise because of superior reactant movement and higher surface to volume proportion (which prompts better ignition). The adding up of nanoparticles brings down the combustion commencement actuation hotness and improves the oxidation of fuel. The most extreme reduction in hydrocarbons discharge for Jatropha biodiesel using nanoparticles (J20T25) was concerning 22% at 75% of engine load compare and precise diesel oil [45].

Researchers study the brake power of three different types of fuel as Tree of heaven, Evening primrose, Camelina biodiesel. It was found that the Tree of heaven, evening primrose biofuel increase concentration of blending with diesel decrease the brake power of the engine. It may due to the difference in the property of biofuel because biofuel properties influence the fuel quality and deliver power. The important properties are heating value, lubricity, and viscosity. Heating value mostly influences the brake power. Many biofuels have a lower heating value than diesel. But Camelina biodiesel has more brake power in comparison to diesel. It may due to an increase in the concentration of biofuel resulting increase in combustion pressure, increased combustion rate, and another factor that may increase the oxygen molecule and proper combustion take place resulting in higher heating value in comparison to other fuel. It was seen that Camelina biodiesel is better fuel in regarding to brake power in comparison with other fuel. When graphene oxide (GO) nanoparticles mix the biofuel. Brake power increases significantly. But maximum power increase was seen in Camelina biofuel. It was found that BSFC also decreases by mixing GO nanoparticles. Maximum reduction was seen in the Camelina biodiesel in comparison with the other two biodiesel blending with nanoparticle GO. Besides, B20GO60 used for Camelina biodiesel has the lowest BSFC. Generally, GO additives will reduce BSFC. This is due to GO nanoparticle additives have a high surface to volume ratio resulting in better combustion occurred inside the combustion chamber [46].

By mixing nanoparticle GO with biodiesel like Camelina, co emissions decreased significantly. Nanoparticle GO provides a large surface area to volume ratio which increases the chemical reactivity of air-fuel mixing resulting in decrease ignition delay. Nanoparticle reduces the heat transfer time into air-fuel mixing and flame propagation inside the air-fuel mixing resulting in ignition delay reduced. An ignition delay period responsible for the proper combustion as these types of CO emissions is reduced. On the other hand, the nanoparticle can convert CO into CO_2 due to which CO emissions decreased but CO_2 increases in comparison to biodiesel fuel without nanoparticle [46–48].

The quantity depends upon the NO_X generation depends on factors like oxygen content inside the combustion chamber, combustion chamber temperature, ignition delay, and compression ratio [35]. It was seen by the researcher that NO_X emissions increased by mixing the biofuel because biofuel contains more O_2 molecule comparatively than diesel. Compared with all dissimilar ratios of biodiesel in diverse fuels, the biodiesel fuel formed by Evening primrose seeds at the ratio of 10% has inferior NO_X emissions [41]. Also it biofuel improves the combustion due to which combustion temperature and cylinder pressure increase resulting in NO_X emissions increases. By mixing GO nanoparticles improve the combustion and increases the cylinder pressure and temperature resulting in NO_X emission increases significantly. It was found that B20 GO 60 for camelina blending has a lesser NO_X emission in comparison to other testing fuels [41, 49].

3 Conclusion

From the writing survey, it tends to be reasoned that because of increment in surface ratio to volume proportion and the expansion in catalyst action of nanoparticles, nanoparticle finds very effective to use as additive substances in diesel in the type of metal and metal oxides. Nano liquids improve the burning because of oxygen donor behavior, higher thermal conductivity, and higher reactive rate. The significant purpose of the extensive survey study can be summed up as follows:

(i) According to the investigation of the survey paper, it was discovered that the expansion of nanoparticles improves the properties of biofuel like increment the heating value, cetane number, viscosity, flash point, and fuel density e of diesel and biodiesel fuel.

(ii) It can be observed from the research paper that the nanofluid upgrades fuel ignition because of the improvement of heat transfer rate, catalyst activity, and air-fuel blending rate. Because of progress in the ignition, increments in calorific value the brake specific consumption of fuel diminishes.

(iii) It is clear from the literature that the utilization of nanofluid added substances increases brake thermal efficiency by promoting total combustion. The outcome shows that expanding the quantitative expansion of nanofluid added substances can improve the thermal efficiency of the brake.

(iv) It was seen from writing that the addition of nanoparticles generally decreased the NO_X because of a higher cetane number, and a decrease in HC because of a higher evaporation rate and catalyst oxidation. Because of the higher dissipation rate and decreased ignition delay, most researchers observed lower flue gas emissions.

(v) It was seen by many authors that there is a decline in CO emissions because of improved combustion after the addition of nanoparticle added substances. At a higher concentration of nanofluid added substances, higher CO discharges were observed. On another angle, it was seen that after the addition of nanoparticles cylinder pressure and heat dissipation increases because of proper combustion and higher evaporation rate of fuel in the burning chamber.

References

1. Kruczyński, S.W.: Performance and emission of CI engine fuelled with camelina sativaoil. Energy Convers. Manag. **65**, 1–6 (2013)
2. Jaichandar, S., Annamalai, K.: Combined impact of injection pressure and combustion chamber geometry on the performance of a biodiesel fueled diesel engine. Energy **55**, 330–339 (2013)
3. Sydborm, A., Blomberg, A., Parnia, S., Stenfors, N., Sandstrom, T., Dahlen, S.E.: Health effects of diesel exhaust emissions. Eur. Respir. J. **17**, 733–46. ISSN 0903-1936 (2001)
4. Patil, M.M., Muhammed, A.M., Appaiah, K.A.A.: Lipids and fatty acid profiling of major Indian garcinia fruit: a comparative study and its nutritional impact. J. Am. Oil Chem. Soc. **93** (6), 823–836 (2016)
5. Aalam, C.S., Saravanan, C.G.: Effects of nano metal oxide blended Mahua biodiesel 539 on CRDI diesel engine. Ain Shams Eng. J. 0–7 (2015)
6. Wang, X., Liu, X., Zhao, C., Ding, Y., Xu, P.: Biodiesel production in packed-bed reactors using lipase-nanoparticle biocomposite. Bioresour. Technol. **102**(10), 542 6352–6355 (2011)
7. Kobayashi, M., Aoyagi, Y., Adachi, T., Murayama, T.: Effective BSFC and NO_X reduction on super clean diesel of heavy duty diesel engine by high boosting and high EGR rate. SAE technical paper 2011-01-0369 (2011)
8. Kiplimo, R., Tomita, E., Kawahara, N., Yokobe, S.: Effects of spray impingement, injection parameters, and EGR on the combustion and emission characteristics of a PCCI diesel engine. Appl. Therm. Eng. **37**, 165–175 (2012)

9. Robinson, K., Ye, S., Yap, Y., Kolaczkowski, S.T.: Application of a methodology to assess the performance of a full-scale diesel oxidation catalyst during cold and hot start NEDC drive cycles. Chem. Eng. Res. Des. **91**, 1292–1306 (2013)
10. Gautam, R., Ansari, N.A., Thakur, P., Sharma, A., Singh, Y.: Status of biofuel in India with production and performance characteristics: a review. ISSN: 0143-0750 (Print) 2162-8246 (Online). Journal homepage: https://www.tandfonline.com/loi/taen20
11. Singh, R.C., Chaudhary, R., Pandey, R.K., Maji, S., Babbar, A., Chauhan, B.S., Gautam, R., Mishra, C.: Performance evaluation of an air cooled diesel engine fuelled with neat neem oil and diesel blends. J. Biofuels **3**(1), 58–64 (2012) https://doi.org/10.5958/j.0976-3015.3.1.006
12. Singh, V., Agarwal, T., Saroha, N., Gautam, R.: Performance emissions and combustion analysis of CI engine using ethyl ester kusum oil and butanol blends. Downloaded from SAE International by Vishal Singh, Tuesday, March 19 (2019)
13. Makwana, N.R., Dabhi, S.K.: Automotive exhaust technology after treatment for the reduction of emission—a review study. Int. J. Eng. Res. Technol. (IJERT) **2**, 876–880 (2013)
14. Sajith, V., Sandhya, M., Sobhan, C.B.: An investigation into the effect of inclusion of cerium oxide nanoparticles on the physicochemical properties of diesel oil. ASME Mater. Div. Publ. —ASME MD 333–8 (2006)
15. Sajeevan, A.C., Sajith, V.: Diesel engine emission reduction using catalytic nanoparticles: an experimental investigation. J. Eng. 1–9 589382 (2013)
16. Soudagar, M.E.M., Nik-Ghazali, N., Abul Kalam, M.D., Badruddin, I.A., Banapurmath, N.R., Akram, N.: The effect of nano-additives in diesel-biodiesel fuel blends: a comprehensive review on stability, engine performance and emission characteristics. Energy Convers. Manage. **178**, 146–77 (2018)
17. Aalam, C.S., Saravanan, C.G.: Effects of nano metal oxide blended Mahua biodiesel on CRDI diesel engine. Ain Shams Eng. J. **8**, 689–696 (2017)
18. Aalam, C.S., Saravanan, C.G., Kannan, M.: Experimental investigations on a CRDI systemassisted diesel engine fuelled with aluminium oxide nanoparticles blended biodiesel. Alexandria Eng. J. **54**, 351–358 (2015)
19. Bhuiyan, M.H.U., Saidur, R., Amalina, M.A., Mostafizur, R.M., Islam, A.: Effect of nanoparticles concentration and their sizes on surface tension of nanofluids. Proc. Eng. **105**, 431–437 (2015)
20. Tanvir, S., Qiao, L.: Surface tension of nanofluid-type fuels containing suspended nanomaterials. Nanoscale Res. Lett. **7**, 1–10 (2012)
21. Prabu, A., Ramachandran, B.A.: Performance, combustion and emission characteristics of a D.I. diesel engine fuelled with nanoparticle blended jatropha biodiesel. Periodica Polytech. Mech. Eng. **59**(2), 88–93 (2015) https://doi.org/10.3311/ppme.7766
22. Syed Aalam, C., Saravanan, C.G., Kannan, M.: Experimental investigations on a CRDI system assisted diesel engine fuelled with aluminum oxide nanoparticles blended biodiesel. Alexandria Eng. J. **54**(3) (2015)
23. Yuvrajan, D., Bheemkumar, N., Ganesan, S.: A comprehensive study on emission and performance characterstics of a diesel engine fueled with nanoparticle-blended biodiesel. Envirn. Sci. Pollut. Res. 11356-019-04446-1 (2018)
24. El-Araby, R., Ashraf Amin.: Study on the characterstics of palm oil biodiesel diesel fuel blend. Egypt. J. Pet. **27**, 187–194 (2018)
25. Sani, F.M., Abdul malik, I.O., Rufai, I.A.: Performance and emission characteristics of compression ignition engines using biodiesel as a fuel: a review. Asian J. Nat. Appl. Sci. **2**(4), 65–72 (2013)
26. Saroj, S.K., Gautam, R., Shubhendu, N.: Reduction of free fatty acid of jatropha oil using Taguchi approach. Int. J. Sci. Eng. Appl. Sci. (IJSEAS) **2**(9), ISSN: 2395-3470 (2016)
27. Prabu, A.: Nanoparticles as additive in biodiesel on the working characteristics of a DI diesel engine. Ain Shams Eng. J. (2017)
28. Ganesh, D., Gowrishankar, G.: Effect of nano-fuel additive on emission reduction in a Biodiesel fuelled CI engine. IEEE 3453–3459 (2011)

29. Selvan, V.A.M., Anand, R.B., Udayakumar, M.: Effects of cerium oxide nanoparticle addition in diesel and diesel-biodiesel-ethanol blends on the performance and emission characteristics of a CI engine. J. Eng. Appl. Sci. **4**, 1–6 (2009)
30. Sajeevan, A.C., Sajith, V.: Diesel engine emission reduction using catalytic nanoparticles: an experimental investigation. J. Eng. 1–9 (2013). https://doi.org/10.1155/2013/589382
31. Sajith, V., Sobhan, C.B., Peterson, G.P.: Experimental investigations on the effects of cerium oxide nanoparticle fuel additives on biodiesel. Adv. Mech. Eng. (2010). https://doi.org/10.1155/2010/581407
32. Gan, Y., Qiao, L.: Evaporation characteristics of fuel droplets with the addition of nanoparticles under natural and forced convections. Int. J. Heat Mass Transf. **54**, 4913–4922 (2011). https://doi.org/10.1016/j.ijheatmasstransfer.2011.07.003
33. Thangavelu, S.K., Ahmed, A.S., Ani, F.N.: Impact of metals on corrosive behavior of biodiesel-diesel-ethanol (BDE) alternative fuel. Renew. Energy **94**, 1–9 (2016). https://doi.org/10.1016/j.renene.2016.03.015
34. Gumus, S., Ozcan, H., Ozbey, M., Topaloglu, B.: Aluminum oxide and copper oxide nanodiesel fuel properties and usage in a compression ignition engine. Fuel **163**, 80–87 (2016). https://doi.org/10.1016/j.fuel.2014.09.008
35. Ahmed, I., El-Seesy, Ali, M.A., Attia, H.M., El-Batsh.: The effect of aluminum oxide nanoparticles addition with Jojoba methyl ester-diesel fuel blend on a diesel engine performance, combustion and emission characteristics. Fuel **224**, 147–166 (2018)
36. Venu, H., Madhavan, V.: Effect of Al_2O_3 nanoparticles in biodiesel-diesel-ethanol blends at various injection strategies: performance, combustion and emission characteristics. Fuel **186**, 176–189 (2016). https://doi.org/10.1016/j.fuel.2016.08.046
37. Agarwal, D., Kumar, L., Agarwal, A.K.: Performance evaluation of a vegetable oil fuelled compression ignition engine. Renew. Energy **33**(6), 1147e56 (2008)
38. Baiju, B., Naik, M.K., Das, L.M.: A comparative evaluation of compression ignitionengine characteristics using methyl and ethyl esters of Karanja oil. Renew. Energy **34**(6), 1616e21 (2009)
39. Jiaqiang, E., Liu, T., Yang, W.M., Li, J., Gong, J., Deng, Y.: Effects of fatty acid methyl esters proportion on combustion and emission characteristics of a biodiesel fueled diesel engine. Energy Convers. Manag. **117**, 410e9 (2016)
40. Shiva, K., Dinesha, P., Marc, A.: Rosen effect of injection pressure on the combustion, performance and emission characteristics of a biodiesel engine with cerium oxide nanoparticle additive. Energy **185**, 1163e1173 (2019)
41. Amith, P.K., Ramesh, R.B.B., Yuvarajan, D.: Emission analysis on the effect of nanoparticles on neat biodiesel in unmodified diesel engine. Environ. Sci. Pollut. Res. https://doi.org/10.1007/s11356-017-9973-6
42. Abdel, R., Gad, M.S., Omar, T.: Effect of aluminum oxide nano-particle in jatropha biodiesel on performance, emissions and combustion characteristics of DI diesel engine. Int. J. Res. Appl. Sci. Eng. Technol. **5**(4), 358–72 (2017)
43. Arockiasamy, P., Anand, R.B.: Performance, combustion and emission characteristics of a D.I. diesel engine fuelled with nanoparticle blended Jatropha biodiesel. Periodica Polytech. Mech. Eng. **59**(2), 88–93 (2015)
44. Sadhik, B.J., Anand, R.B.: Effects of nanoparticle additive in the water diesel emulsion fuel on the performance, emission and combustion characteristics of a diesel engine. J. Vehicle Design **59**(2/3), 164–181 (2012)
45. Gada, M.S., Jayarajb, S.: A comparative study on the effect of nano-additives on the performance and emissions of a diesel engine run on Jatropha biodiesel. Fuel **267**, (2020)
46. Hoseini, S.S., Najafi, G.B., Ghobadian, M.T., Ebadi, R., Mamat, T., Yusaf, T.: Biodiesels from three feedstock: the effect of graphene oxide (GO) nanoparticles diesel engine parameters fuelled with biodiesel. Renew. Energy **145**, 190e201 (2020)
47. Prabu, A., Anand, R.B.: Emission control strategy by adding alumina and cerium oxide nano particle in biodiesel, J. Energy Inst. **89**(3), 366e372 (2016)

48. Self, W., Seal, S., Dowding, J.: Methods for scavenging nitric oxide using cerium oxide nanoparticles. Google Patents (2014)
49. Muralidharan, K., Vasudevan, D.: Performance, emission and combustion characteristics of a variable compression ratio engine using methyl esters of waste cooking oil and diesel blends, Appl. Energy **88**(11), 3959e3968 (2011)

Experimental and Computational Analysis of Household Cook Stoves: A Review

Anoop Singh, Vinod Kumar Yadav, Amardeep, Mahesh Kumar Maddheshiya, Sagar Sharma, Manav Jha, and Prajjwal Singh

1 Introduction

In India, LPG connections that have been distributed among the people lying below poverty line under the PMUY was around Eighty million in the year 2016 [1]. In most of the developing nations, liquefied petroleum gas (LPG) is commonly used as fuel in household cooking appliances. LPG cooking stoves contribute essentially to indoor air pollution [1]. Developing country like India has consumed 24.9 million tons LPG during 2018–2019. It is 6.9% [2] higher than previous year. The lower thermal efficiency and emission of pollutant gases causes indoor air pollution. So, there is energetic need to explore the approaches to improve the efficiency of LPG cooking stoves. Berry et al. [3] worked on the design of gas burners and reported about the advanced methods of testing and arrangement. They also studied about the ports of the burners, injecting tube design, principles of rate of injection of air into the burner and flow of gas through different orifices design. The rate of discharge of gas orifices found to vary greatly with a variation of the approach angle and length of channel or tube of the orifice. They applied various strategies for testing the performance of the burner. Designing and optimization of various injection tubes were studied by changing the position of orifice burner used to get expected result. The position of orifice was also changed for getting good results. Tamir et al. [4] found a new gas burner design generating a swirling flame. Figure 1 shows the Swirling flame type LPG gas burner. This burner is 10–30% more efficient than the conventional burner. In Swirling effect, due to high-shear stress developed due to rotating moment, the combustion intensity gets increased. They explored the effect of the design parameters under different working conditions for normal and synthetic gases.

A. Singh · V. K. Yadav (✉) · Amardeep · M. K. Maddheshiya · S. Sharma · M. Jha · P. Singh
Department of Mechanical Engineering, GL Bajaj Institute of Technology and Management, Greater Noida 201306, India
e-mail: vinod.yadav@glbitm.ac.in

L. M. Das et al. (eds.), *Recent Trends in Thermal Engineering*,
Lecture Notes in Mechanical Engineering,
https://doi.org/10.1007/978-981-16-3428-4_8

Pandey et al. [6] surveyed in the high-altitude areas (Nepal) to know the success of smokeless ovens in dropping indoor air pollution. In twenty households, the personal exposure levels and the harmful suspended particulates were monitored. Gravimetric analysis with the filters in traditional stoves and dragger tube analyses were conducted. In the survey they found that traditional stoves emit more smoke compared to smoke stoves. Kandpal et al. [7] experimentally investigated the indoor air contamination induced by local cooking appliances, that utilizes coal, kerosene and LPG. They found that the main reason for indoor air pollution was the use of unvented coal, kerosene and LPG which accumulate high concentration of impurities in the indoor environment while cooking. It was reported that the well-ventilated kitchens must be used to reduce the health impacts of the pollutants. Weinberg [8] used premixing flame for pre-heating of incoming LPG mixture using solid porous medium. PMC is physically and optically very active in heat transfer property and hence used in various application like radiant burner [9], heat exchanger [10], heating [11], gaseous turbine [12], hydrogen production [13] and ignition compression engine [14]. Jugjai and Sanitjai [15] presented a new design to enhance the mostly used conventional open-flame atmospheric burner. Efficiency of the burner was improved using the porous medium technology by constructing an internal heat recirculation system to capture and recirculate some of the enthalpy of the exhaust gases. When a porous medium of an appropriate optical thickness is used, the PM (porous medium) changes the enthalpy of the exhaust smokes to thermal radiation which is fed back to burner inlet. Due to the internal heat recirculation, the efficiency of the PRRB is approximately 10% higher than the standard burner (SB) and PRB. Jugjai and Rungsimuntuchat [16] used porous material (PM) in LPG burner for improving the thermal efficiency of burners. Experimental studies showed that using PM, thermal efficiency can be enhanced approximately by 20% with an energy saving of 30%. D'Sa and Murthy [17] showed the importance of LPG against biomass-based fuels in their article. They reported that out of whole indian population, only 17.5% depends on LPG at present. They analyzed the factor affecting current demand and future scenarios to increase the household use of LPG. The problems regarding affordability, pricing and reliable distribution are the main barriers. So, they suggested the policies by recognizing the challenges to overcome the reported problems. Hou et al. [18] studied and experimented the thermal efficiency and CO released using swirl and radial type burner and explored effect of heat input, primary mixing of air and

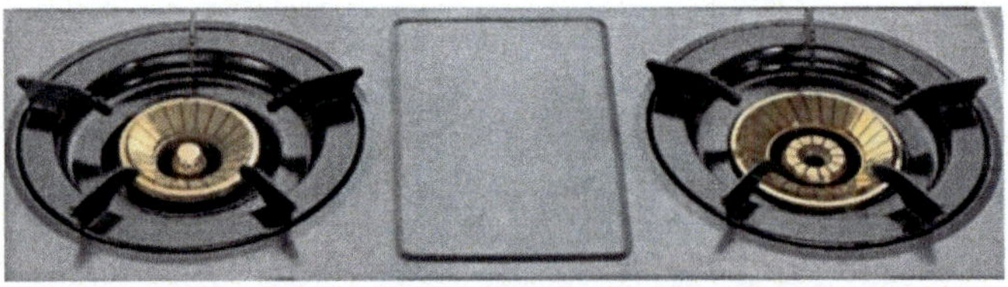

Fig. 1 Design of Swirling flame type LPG gas burners [5]

loading height. Pantangi et al. [19] suggested to work on porous medium combustion instead of conventional burner by using different types of ball bearing which yielded increase in efficiency about 73% with 10% cost saving. They performed their experiment on three different combinations of porous media. By using mild steel chips in the combustion zone and pebbles in preheating zone, the maximum thermal efficiency was 73%. Tierney et al. [20] described the fluid and solid porous medium cooperation. Porous medium helped in improving efficiency to overcome pollutant emission. They also provide stability in operating power range and can be operated within lower flammability limits. Validation of chemical mechanism, used in numerical model is being confirmed by CHEMKIN 'reaction design 2006'. Validation is conducted using a CFD fluid temperature on a plug flow as output keeping same inlet conditions. Panting et al. [21] worked on the porous radiant burners (PRB) used for LPG cooking gas. They reported that by using porous medium, which is made of two layers, that consists of Alumina and Silica Carbide, the thermal efficiency is high. Bhatt et al. [22] altered the pan support and conventional design of the burner of a domestic LPG to improve the thermal efficiency. The old one was 13% less efficient against minimum thermal efficiency (64%) by the BIS. By analyzing the key main factors of combustion and heat transfer, they found that the thermal efficiency of LPG cook stoves can be increased with the help of swirling flames, porous inserts, and low-thermal inertia pan. Swirling flame burner cap and pan support is the best modification among all the three options. They also found that with an optimum pan height, the thermal efficiency gets boosted by 2%, which is sufficient to save gas worth of 361.677 crores as per the rate of 20 Rs./kg on consumption of 5.56 MT annually. Mishra et al. [23] studied about the Porous Radiant Burners made of SiC. Different heat inputs (5–10 kW) effects were studied. For LPG burner, the thermal efficiency ranged from 30 to 40%, 350 to 1145 ppm CO, and 40 to 109 ppm NO_x. It has a maximum efficiency of 50%, which is more than the CS (conventional stoves). To overcome difficulty in free flame combustion, porous medium combustion (PMC) was developed. PMC has better power density, high power dynamic reach and emit very less NO and CO emissions. Thermal efficiency of PRB is 40–50%. By increasing power intensity, the efficiency of PRB gets decreased. PRB gives maximum thermal efficiency at the load of 10 kW. Muthukumar et al. [24] presented new design of PRB (porous radiant burner) for LPG domestic cooking application. Using water boiling test (BIS: 4346:2002), the thermal efficiency of burner was found. To study the effect of porosity, the burners were tested at 80, 85 and 90% porosities at comparable ratios and wattages. The highest thermal efficiency with reduced emissions was found 90% porosity. Khan and Saxena [25] studied the effects of different designs of burner heads on performance on LPG cook stoves and reported that the burners made up of brass were more effective than the iron made burners. In addition, the burners having flat and flower face are more effective than the regular burners, since the flat face brass burner has thermal efficiency of 58%, while the regular burner reached up to 48% efficiency only. This process of changing burner heads was very simple. Boggavarapu at el. [26] worked experimentally and numerically on LPG and PNG and measured the thermal efficiency of a conventional burner. Three dimensional CFD (computational fluid dynamic) modeling of the steady-state flow, ignition and heat flow to the pot is reported. They modified the circular hole of burner and also worked on the loading height. Khan et al. [27] concluded that the efficiency of the cooking system can be increases by using best design burners, using pots with plane bottoms, working at flow rates that deliver the maximum efficiency, and using a porous burner. Porous media combustion provides a 3-d matrix mixing chamber to achieve a higher value of thermal conductivity and emissivity. Aroonjarattham [28] worked with outer and inner rings of the burner heads for number of holes and hole angles to calculate its effect on thermal efficiency. Terracciano et al. [29] designed a bi-fuel heterogeneous combustor containing silicon carbide ceramic media in the combustion chamber. The combustor was operated for both lean and rich fuel-air mixtures. The flow loop of combustor consists of 4 inter-connected devices. The flow loop combustor has mixing chamber, heat exchanger, vaporization chamber for the injection of fuel and the temperature was measured with the help of thermo couples at various locations. It was observed that the capacity of combustor, while working with gaseous methane and air, can be maintained even at a small equivalence ratio of 0.50. Makmool et al. [30] developed a novel self-aspirating liquid fuel annular porous medium burner (SLAPMB).

Laphirattanakul and Charoensuk [31] placed a needle inside the fuel nozzle and achieved the refinement of air flow rate for the LPG premixed burner. The results shows that with narrow design of the jet pattern, the degree of penetration will be higher. The momentum decay rate was 48% at z/D = 30 for lower RSS (Reynolds shearing stress) related to jet shape upstream. Near the exit plane, the main reason for the flow of jet momentum through the CR (central region) of the jet was due to the correlation with the vorticity range of the internal core. Firing rate between 1.8 and 4.4 kW obtained the best range for air entrainment at 25%. Mishra and Muthu Kumar [32] worked on the design of a self-aspirating two-layer PRB for LPG cooking stove. The performances of the LPG stove with PRB have been tested by following the rules described in BIS (Bureau of Indian Standards) 4246:2002. In the power range of 1–3 kW, the maximum thermal efficiency of the CB and PRB is found as 65% and 75.1% respectively. The emission of NO_x and CO ranges between 0.2–3.5 ppm and 30–140 ppm respectively. The NO_x and CO emissions from domestic LPG cooking stoves (1–3 kW) ranges between 5–25 ppm and 220–550 ppm respectively. Figure 2a, b shows that the newly designed PRB is self-aspirated, so it works on the natural draft without any safety issues.

Bantu et al. [33] developed an efficient charcoal stove by applying high-density rock and heat retaining methods. The objective was to reduce the heat losses by retaining the heat. The gas stove design assimilated the interplay of physical and thermal qualities of granite rocks with heat loss theory to give a thermal efficient element. Through detailed literature survey, it was found that the granite rock produces high-thermal storage characteristics. Fakinle et al. [34] calculated the efficiencies of the different types of burners used in many countries. They analyzed their results and compared the thermal efficiencies of the different versions of the burners. They also compared the emissions coming out of different burners. Sener

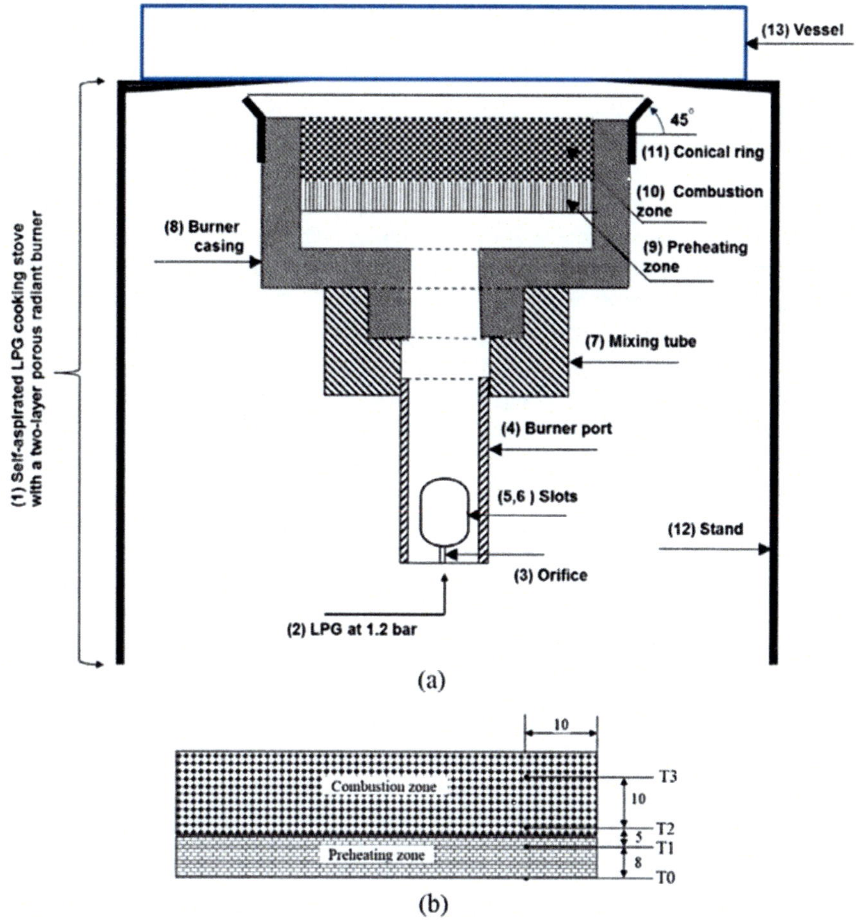

Fig. 2 **a** Schematic diagram of the newly developed PRB [47] and **b** location of thermocouple at different axial

et al. [35] observed geometrical parameters in domestic burner with crescent flame at optimal temperature distribution and thermal efficiency. In thier study, a burner containing 24 channels is fixed vertically with crescent flame around the cooker. They reported that by heating the surface, the crescent flame channels gets homogeneously distributed.

CAD software is used to create the geometry of the burner. To analyze the temperature distribution on the surface of the cooker and emission values, they experimented over three different cookers with bottom wall distance from burner head of 5, 10 and 15 cm with three different diameters (25, 30, 35 cm). Lowest surface of the cooker and unburnt HC, CO and NO emissions were observed and the thermal efficiency was calculated. They reported that by increasing the cooker diameter and distance, the temperature on the surface and the emission values of unburnt CO, NO, HC decreases. Fedak et al. [36] experimented and investigated the air pollution induced by the solid fuel used in the household that affect the human health and environment. They performed a start-up process to understand the cook stove emission and recorded the data of the type of fuel used to start the cook stove globally. Wichangarm et al. [37] worked or the prediction of the thermal efficiency of LPG burner using computational fluid dynamics data. They found a good agreement between the computational and experimental data for flow velocities and the combustion temperature distribution. The correlated data which is obtained from the both analysis is used for the calculation of the thermal efficiency. Sutar et al. [38] reported about the common cooking procedures with the best combination of pot size. They experimented for a 4-member family, based on the survey of 80 households. They used three settings of the flame (High, medium and low) and four types of pots for smaller burner of domestic LPG stove. Bakry et al. [39] started at low powers and studied dormant problems that come across while combustion in burners. They obserevd that correct combustion process could not occur near the downstream surfaces. Two patterns of combustion were detected based on air-ratio. For differentiating between the values which were above or below the limit, they used suspended free flame and explosive flame plume, respectively. The Sic—foam and two region PIM burners were used as supercritical region to find out the variation of the temperature. This was done with LPG at low pressure (0.576–2.535) and low powers (0.73–1.58). For a smooth and adequate start, limiting power (depending on PIM configuration) was proposed. They concluded that the developers should acknowledge these restrictions on power and values for the PIM designed by them, prior to their implementation in real appliances.

Kephart et al. [40] inspected the Peruvian Andes to distinguish concentrations and personal exposures to NO_2 under a randomized controlled trial. They conducted trial of biomass with LPG intervention between 50 interference members and 50 control members of high-temporal resolution homes and computed the NO_2 level in the kitchen area. Shen et al. [41] performed test on different type of LPG gas burner, which are used in Japan and China market for calculating the thermal efficiency and emissivity of the LPG gas. They also performed the test on used old burner. They compared all data and calculated the average. Dhable et al. [42] improved the efficiency of domestic LPG burner by changing the angle of fuel holes by creating the swirl in the burner. The efficiency yield was 60%, which in turn, is twice as compared to conventional burner. The experiments were conducted to study the impinging flame shapes, temperature distribution and to improve the heating efficiency of domestic LPG burner. Trimis and Durst [43] described the combustion in porous media by designing combustor-heat exchanger. For experimental work tubes, filled with PM and premixed fuel-air mixture, was used. From result analysis, they found that the PM burner has excellent power dynamic range (20:1). Availability of ceramic material and their heat transfer properties elaborated the scope for improvement and excellent emission values of burners. Between a wide range of gas properties and air-fuel ratios, a stable combustion was ensured. It was observed that systems may be indifferent in varying the gas mixture properties.

Laphirattanakul et al. [44] investigated the flames of modified premixed LPG burner for ceramics industry with dissimilar density of porous media within the range of 1–5 L/min. By maintaining cone shaped flames, which was necessary for this application, ignition took place inside the porous domain and propagated downstream. With the port density of 15 and 20 ppi, made of alumina porous foams, were constructed in cylindrical geometry and replaced for an iron casted conventional burner. It was observed that with this unique design, the secondary air could be induced naturally. Flame visibility was observed in addition to the gas mixture temperature profile at different fuel-air mixture ratio prior to combustion. It was found that variation in pore density can significantly change the combustion characteristic. Lucky et al. [45] Found that in Bangladesh two most commonly used fuel are biomass and natural gas, out of which biomass is used widely. Eventually, it will take time to shift towards the fuels like natural gas and LPG. In the cooking system, the energy consumption which increases or decreases the efficiency, depends upon the stove design, material and regulation of fuel. The natural gas cook stoves, which is conventional type, is widely use in Bangladesh. Very less modified burners have been developed and still few of them are available in the market. Pan gives good efficiency at all power values in perforated head stoves as well as conventional stoves. Perforated head cook stoves gave higher efficiencies at different gas flows, when their shields were made up of sheet metal. At the time of construction, smaller sides fixed together using pair of nut and bolt on any one exit hole of the flue gases for achieving variation in circular shield diameter. For determining optimum diameter of shield perforated head burners were used. The maximum efficiency of pot and pan containing shields were 57.8% (without shield is 51.5) and 62.2% (without shield is 59.1), respectively. Anozieaet al. [46] surveyed in Nigeria and they studied different factors for cooking like energy costs, efficiency, air pollution, energy utilization and the influence of the energy policy. They carried out the survey to find the cooking energy, use pattern in the city and village areas by water boiling and cooking experiment with the help of various energy sources. They reported that wood is less expensive than LPG. For boiling water, electric hot plate was reported as the highly efficient method. The pollutants in air showed that fuel wood is the predominant source. They also observed that the energy policy had made no effect in the cooking energy area. Recommendations were applied for improving the energy supply to remove the obstructions that forbid the implementation of the recommendation.

Thacker et al. [48] studied the ways to improve the technical capabilities (performance efficiency and reduction of emissions) of the cookstoves. Their design of pot skirt didn't achieved the same level of performance as that of the advanced cookstoves. Goldenberg et al. [49] reported that the developing countries effectively manage climate issues and are also being supported politically. Biomass cook smoke causes millions (2.2–3.6) of deaths yearly. Utilization of biomass cook stove, produces CO_2, CH_4, NO, CO and various other harmful gases [51]. The way of clean energy is the combined package of climate, health and energy. Therefore, it is important for all developing countries to balance all these parameters in parallel. There is a need to create most effective solutions capable of eliminating all these problems and create a path toward sustainability. LPG is a rare fuel having matching qualities like clean combustion, effective burning and easy storage [50].

2 Review of Experimental Setup Used in Thermal Efficiency Calculation

In this section, the ignition of LPG with constant mass, over a locally available burner, is discussed. Duration of the gas was varied over a certain period of time to allow and change the inlet pressure of the gas to the burner head [34]. There are different authors who have used different formula for calculating the thermal efficiency. Most of them use only two formulae Fakinle et al. [34] and Anozie et al. [46] used Eq. 1 for calculating the burner's thermal energy. Figure 3: Shows the experimental set-up for efficiency calculation.

Thermal efficiency of the stove is given as follows:

$$\eta_{th} = \frac{(\mathrm{Ww} * \mathrm{Cw}) * (T2 - T1) + (\mathrm{Me} * L)}{(\mathrm{Wf}) * \mathrm{CV}} \tag{1}$$

where,

Ww = Weight of the water (in kg),
Me = Mass of the water evaporated (kg),
Cw = Specific heat of the water (kJ/kg-K),
Cv = Specific heat of the pot (kJ/kg-K),
Wf = Weight of the fuel burnt (kg),
CV = Calorific value of the test fuel (kJ/kg).

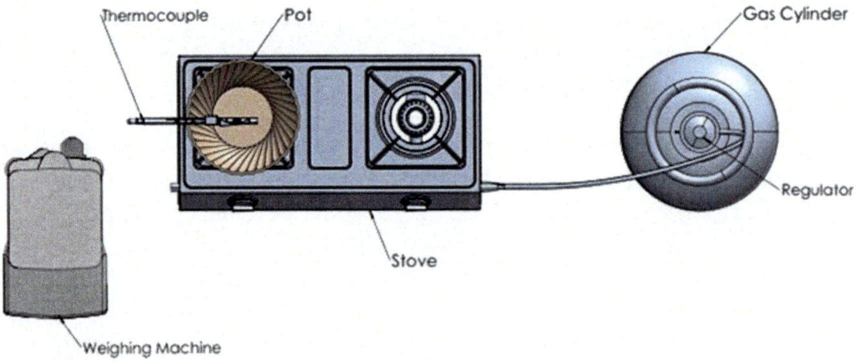

Fig. 3 Experimental set-up for efficiency calculation

Table 1 Summary of literatures based on burner material and type

Material used in the burner	Type of Burner	CO_2 (Carbon dioxide)	CO (Carbon monoxide)	CH_4 (Methane)	NO_x (Nitrogen oxide)	Thermal efficiency (%)	Country	Ref.
SiC-Combustion zone Al_2O_3 -Pre heating zone	PRB, 90% porosity	–	(12–124) mg/m^3	–	(0–0.75) mg/m^3	75	India	[24]
Brass	Flat and flower face brass burner of weight 0.5 kg	–	–	–	–	58	India	[25]
Copper alloy	Conventional burner	–	900 PPM	–	50 PPM	33.6	Thailand	[30]
SiC-Combustion zone Al_2O_3-Pre heating zone	PRB, 3PRB Internal Dia.-17,19,21 mm Orifice Dia.-0.25,0.35,0.5 mm	–	(30–80) PPM	–	(0.2–2.2) PPM	75.1	India	[32]
Cast iron	Number of Port = 285 Indian Burner diameter of port = 1 mm	1792 ± 0 mg/m^3	48.5 ± 26.5 mg/m^3	–	2.5 ± 64.5 mg/m^3	70	Nigeria	[34]
Copper alloy	Conventional burner	142 ± 17 mg/MJ	0.77 ± 0.55 mg/MJ	5.6 ± 8.2 mg/MJ	46 ± 9 mg/MJ	51 ± 6	China and Japan	[41]

Pantangi et al. [19] and Khan et al. [27] used Eq. 2 for calculating the burner's thermal efficiency.

Thermal efficiency of the stove is given as follows:

$$\eta_{th} \frac{(\mathrm{Ww} * \mathrm{Cw} + \mathrm{Wv} * \mathrm{Cv}) * (T2 - T1)}{(\mathrm{Wf}) * \mathrm{CV}} \tag{2}$$

where,

Ww = Weight of the water (kg),
Wv = Weight of the pot (kg),
Cw = Specific heat of the water (kJ/kg-K),
Cv = Specific heat of the pot (kJ/kg-K),
Wf = Weight of the fuel burnt (kg),
CV = Calorific value of the test fuel (kJ/kg)

The literatures dealing with these calulations are presented in Table 1.

3 Conclusion

This paper summarized the outcomes of the studies conducted to estimate the performance and efficiency of the stoves used in household applications. Comparative performance using fuels like LPG, kerosene and biogas were also studied. It has been observed that the performance of the burners can be enhanced by changing in burner specification, design and material. Air preheating effect may also play key role for the betterment of cooking stove burners. Many researchers conducted experimental and numerical procedures and reported that the porous burners are better than the conventional burners with enhanced efficiency and reduced emissions. The burner material and design may also play a major role in enhancing the the efficiency of cooking stoves as burners made up of brass were more effective than those which are made up of iron. In addition, the burners with flat face were more effective than the flower faced burners.

The major conclusion drawn from the detailed lietarure review are as follows:

1. The thermal efficiency of conventional burners lies in the range of 33.6–58%.
2. The thermal efficiency of porous radiant burner burners lies in the range of 70–75.1%.

References

1. Mortality from Household Air Pollution-Summary of Results, World Health Organization (2014). https://www.who.int/phe/health_topics/outdoorair/databases/en/
2. https://www.livemint.com/industry/energy/india-s-lpg-use-to-surge-from-record-as-govt-promotes-cleaner-fuel-1556861784697.html. Accessed on Nov 2020
3. Berry, W.M., Brumbaugh, I.V., Moulton, G.F.M., Shawn, G.B.: Design of Atmospheric Burners. Technologic papers of the bureau of standards, September 6 (1921)
4. Tamir, A., Elperin, I., Yotzer, S.: Performance characteristics of a gas burner with A swirling central flame. Energy **14**(7), 373–382 (1989)
5. https://kensoko.com/desc_images/5197_Artboard%202.jpg. Accessed on 24 Dec 2020
6. Pandey, M.R., Nuepane, R.P., Gautam, A., Shreshta, I.B.: The effectiveness of smokeless stoves in reducing indoor air pollution in a rural hill region of Nepal. Mounlain Res Deoe Lopmen **10**(4), 313–320 (1990)
7. Kandpal, J.B., Maheshwarp, R.C., Kandpal, T.C.: Indoor air pollution from domestic cookstoves using coal, kerosene and LPG (1994)
8. Weinberg, F.J.: Combustion temperature: the future. Nature **233**, 239–241 (1971)
9. Qiu, K., Hayden, A.C.S.: Increasing the efficiency of radiant burners by using polymer membranes. Appl. Energy **86**, 349–354 (2009)
10. Delalic, N., Mulahasanovic, DZ., Ganic, E.N.: Porous media compact heat exchanger unit—experiment and analysis. Exp. Thermal Fluid Sci. **28**(2–3), 185–192, ISSN 0894-1777, (2004). https://doi.org/10.1016/S0894-1777(03)00038-4
11. Avdic, F., Adzic, M., Durst, F.: Small scale porous medium combustion system for heat production in households. Appl. Energy **87**, 2148–2155 (2010)
12. Tanaka, R., Shinoda, M., Arai, N.: Combustion characteristics of a heat recirculating ceramic burner using low-calorific fuel. Energy Converse. Manage. **42**, 1897–1907 (2001)
13. Mjaanes, H.P., Chan, L., Mastorakos, E.: Hydrogen production from rich combustion in porous media. Int. J. Hydrogen Energy **30**, 579–592 (2005)
14. Weclas, M.: Porous media in internal combustion engines. In: Cellular Ceramics-Structure, Manufacturing, Properties and Applications. Wiley-VCH Publication (2005)
15. Jugjai, S., Sanitjai, S.: Parametric study of thermal efficiency in a proposed porous radiant recirculated burner: a design concept of future (1996)
16. Jugjai, S., Rungsimuntuchart, N.: High efficiency heat-recirculating domestic gas burners. Exp. Therm. Fluid Sci. **26**, 581–592 (2002)
17. D'Sa, A., Murthy, K.V.N.: LPG as a cooking fuel option for India. Energy Sustain. Dev. **8**(3) (2004)
18. Hou, S.S., Lee, C.Y., Lin, T.H.: Efficiency and emissions of a new domestic gas burner with a swirling flame. Energy Convers. Manag. **48**, 1401–1410 (2007)
19. Pantangi, V.K., Kumar, A.K., Mishra, S.C., Sahoo, N.: Performance analysis of domestic LPG cooking stoves with porous media. Int. Energy J. **8**, 139–144 (2007)
20. Tierney, C., Wood, S., Harris, A.T., Fletcher, D.F.: Computational fluid dynamics modeling of porous burners. Seventh international Conference on CFD in the Minerals and Process Industries CSIRO, Melbourne, Australia 9–11 December
21. Pantangi, V.K., Mishra, S.C., Muthukumar, P., Reddy, R.: Studies on porous radiant burners for LPG (liquefied petroleum gas) cooking applications. Energy **36**, 6074–6080 (2011)
22. Bhatt, D.S., Mulla, I.A., Dhamangaonkar, P.R.: Design and development of burner ring to enhance thermal efficiency of domestic lpg stove and assessment of its performance. 10–12, 385–390 (2011)
23. Mishra, N.K., Muthukumar, P., Mishra, S.C.: Performance tests on medium-scale porous radiant burners for LPG cooking applications. Int. J. Emerg. Technol. Adv. Eng. **3**, 126–130 (2013)
24. Muthukumar, P., Shyma Kumar, P.: Development of novel porous radiant burners for LPG cooking applications. Fuel **23**, 562–566 (2013)

25. Khan, M.Y., Saxena, A.: Performance of LPG cooking stove using different design of burner head. IJERT **2**, ISSN: 2278-018 (2013)
26. Boggavarapu, P., Ray, B., Ravikrishna, R.V.: Thermal efficiency of LPG and PNG-fired burners: experimental and numerical studies. Fuel **116**, 709–715 (2014)
27. Khan, M.Y., Saxena, A., Katiyar, J.: Performance of LPG burner with different size of ball bearing as a porous medium. J. Basic Appl. Eng. Res. **2**
28. Aroonjarattham, P.: The parametric studied of high-pressure gas burner affect thermal efficiency. Eng. J. **20**(33), 86 (2016)
29. Terracciano, A.C., Vasu, S.S., Orlovskaya, N.: Design and development of a porous heterogeneous combustor for efficient heat production by combustion of liquid and gaseous fuels. Appl. Energy **179**, 228–236 (2016)
30. Makmool, U., Pontree, K., Jugjai, S.: Design and development of a novel self-aspirating liquid fuel annular porous medium burner (SLAPMB). Fuel **192**, 83–92 (2017)
31. Laphirattanakul, P., Charoensuk, J.: Effect of central cone-shaped bluff body on performance of premixed LPG burner. Appl. Therm. Eng. **114**, 98–109 (2017)
32. Mishra, N.K., Muthukumar, P.: Development and testing of energy efficient and environment friendly porous radian burner operating on liquified petroleum gas. Appl. Therm. Eng. **129**, 482–489 (2018)
33. Bantu, A.A., Nuwagaba, G., Kizza, S., Turinayo, Y.K.: Design of an improved cooking stove using high density heated rocks and heat retaining techniques. Hindawi J. Renew. Energy 9620103 (2018)
34. Fakinle, B.S., Oke, O.D., Odunalmi, O.A., Sonibare, J.A., Akeredolu, F.A., Oni, O.S.: Emission characterization and performance of conventional liquified petroleum gas cookstoves burner. Cogent Eng. **6**, 1652228 (2019)
35. Şener, R.,Özdemir, M.R., Yangaz, M.U.: Effect of the geometrical parametersina domestic burner with crescent flame channels for an optimal temperature distribution and thermal efficiency. J. Therm. Eng. **5**(6)(10), 171–183 (2019)
36. Fedak, K.M., Good, N., Walker, E., Clark, M., L'Orange, C., Volckens, J., Peel, J.: An expert survey onthe material types used to start cookstoves. Energy. Sustain. Dev. **48**, 59–66 (2019)
37. Wichangarm, M., Matthujak, A., Sriveerakul, Sucharitpwatskul, S., Phongthanapanich, S.: Investigation on thermal efficiency of LPG cooking burner using computational fluid dynamics. Energy **203**, 117849 (2020)
38. Sutar, K.B., Kumar, M., Patel, M.K., Kumar, A., Mokashi, S.R.: Experimental investigation on pot design and efficiency of LPG utilization for some domestic cooking processes. Energy Sustain. Dev. **56**, 67–72 (2020)
39. Bakry, A.I., Rabea, K., El-Fakharany, M.: Starting up implication of the two-region porous inert medium (PIM) burners. Energy **201**, (2020)
40. Kephart, J.L., Fandino-Del-Rio, M., Williams, K.N., Malpartida, G., Lee, A., Steenland, K., Naeher, L.P., Gonzalesi, G.F., Chiang, M., Checkley, W., Koehler, K.: Nitrogen dioxide exposures from LPG stoves in a cleaner-cooking intervention trial. Environ. Int. **146**, (2021)
41. Shen, G., Hays, M.D., Smith, K.R., Williams, C., Fair Cloth, J.W., Jetter, J.J.: Evaluating the performance of household liquefied petroleum gas cookstoves. Environ. Sci. Technol.
42. Dhabale, S., Jaganur, A., Pujeri, G., Dashamani, R.: Improvements in domestic LPG burner
43. Trimis, D., Durst, F.: Combustion in a porous medium-advances and applications 153–168 (2007). https://doi.org/10.1080/00102209608935592
44. Laphirattanakula, P., Laphirattanakulb, A., Charoensuka, J.: Effect of self-entrainment and porous geometry on stability of premixed LPG porous burner. Ate 7947, S1359-4311(16) 30372-6 (2016)
45. Lucky, R.A., Hossain, I.: Efficiency study of bangladeshi cookstoves with an emphasis on gas cookstoves. Energy **26**, 221–237 (2001)
46. Anoziea, A.N., Bakarea, A.R., Sonibarea, J.A., Oyebisi, T.O.: Evaluation of cooking energy cost, efficiency, impact on air pollution and policy in Nigeria. Energy **32**, 1283–1290 (2007)
47. Mishra, S.C., Muthukumar, P., Mishra, N.K.: Self-aspirated LPG domestic cookingstove with a two-layer porous radiant burner. Indian Patent No: 543/KOL (2015)

48. Thacker, K.S., Barger, K.M., Mattson, C.A.: Balancing technical and user objectives in the redesign of aperuviancookstove. Dev. Eng. **2**, 12–19 (2017)
49. Goldemberg, J., Gomez, J.M., Sagar, A., Smith, R.K.: Household air pollution, health, and climate change: cleaning the air. Environ. Res. Lett. **13**, 030201 (2018)
50. Jewitta, S., Atagherb, P., Cliffordb, M.: We cannot stop cooking. Stove stacking, seasonality and the risky practices of household cookstove transitions in Nigeria. Energy Res. Soc. Sci. **61**, 101340 (2020)
51. Teotia, S., Yadav, V.K., Sharma, S., Yadav, J.P.: Effect of porosity and loading height on the performance of house hold LPG gas stoves. J. Process. Mech. Eng. 1–8 IMechE (2021). https://doi.org/10.1177/0954408920987024

Development of Heat Storage Device Assisted with Heat Pipe

Virendra Kumar Yadav, Ambuj Kumar, and Ritesh Jaiswal

1 Introduction

In India, 57.3% of total electric power is generated by using coal, 7.2% by burning of natural gas, 2.0% by nuclear reactor and 0.2% of electricity is generated by using diesel. These non-renewable sources of energy generate a lot of pollution and affect the environment. It is estimated that till 2035, 42 billion ton of CO_2 will be emitted to the atmosphere [1]. To control the pollution, we need to avoid the use of fossil fuels for energy generation and use the renewable source of energy like solar, wind, hydro power plant, etc. The solar energy is available in abundance and it is cheap source of energy. It is a challenging task to store the solar energy in batteries since it is economically costlier [2]. There are two methods to generate power from sun energy (i) Photovoltaic system, (ii) Thermal energy storage (TES) system. The thermal energy storage system uses phase change material (PCM) to store the solar energy which can be used to drive the steam turbine for electricity generation. TES system is attracting the researcher's attention since phase change material can store large amount of energy. Rathod et al. [3] performed experimental study of PCM based thermal storage tank filled with $CaCl_2 \cdot 6H_2O$ (calcium chloride Hexa-hydrate) and Sodium Carbonate Deca-hydrate ($Na_2CO_3 \cdot 10H_2O$). They found that heat storing and releasing capacity of storage tank depends upon flow rate of heat transfer fluid in the stream tube. Sarbu et al. [4] told that heat storing capacity of tank can be improved by using phase change material inside it. Kenisarin et al. [5] used salt hydrates as PCM. It was found that for temperature range of −50–120 °C, salt hydrates act as a good energy storing material. Zhang et al. [6] worked on the design of storage tank. They found that copper tube placement inside the storage tank improve the uniform heat distribution which leads to store more amount of energy. Shabgard et al. [7] investigated the response of TES system using latent

V. K. Yadav (✉) · A. Kumar · R. Jaiswal
Department of Mechanical Engineering, Kamla Nehru Institute of Technology, Sultanpur, Uttar Pradesh, India

L. M. Das et al. (eds.), *Recent Trends in Thermal Engineering*,
Lecture Notes in Mechanical Engineering,
https://doi.org/10.1007/978-981-16-3428-4_9

heat phase change material associated heat exchanger. They concluded that temperature drop across the tank is depends on the number of heat pipe installed into the storage tank. Less number of heat pipe results in high temperature drop across the storage tank and high number of heat pipe results in the lowest temperature drop in the TES system.

1.1 Categories of Energy Storage Method

There are three methods available to store the energy which is as follows:

- Sensible heat based energy storage system: It is the simplest approach to store energy by heating solid or liquid storage medium like sand, molten salt, rocks and water. These medium are economically cheap and non-toxic in nature [3].
- Latent heat-based energy storage system: Latent heat energy storage material is well-recognized as phase change material like paraffin. These material store and release energy by changing their phase. These materials have very high-heat storing capacity. The latent heat material should have high-energy storage density, good thermal conductivity, and chemical stability, non-toxic, inflammable and less hazardous to environment [3].
- Chemical heat-based energy storage system: These energy storage systems include the batteries and which the potential is developed due to chemical reaction of solution [8].

1.2 Phase Change Material

A phase change is material is a substance which absorb or release sufficient amount of energy at phase transition to provide useful heating and cooling. The classification of phase change material is shown in Fig. 1.

1.2.1 Paraffin

The most abundant organic phase change material present is paraffin. Their chemical composition is generally *n*-alkanes. The paraffin PCM has good thermo-physical properties, good nucleation rate and segregation on cooling [9]. These are non-corrosive and chemically stable. Some paraffin PCM generally used is n-Triacontane, n-Eicosane ($C_{20}H_{42}$), Polyethylene, n-Pentadecane, etc.

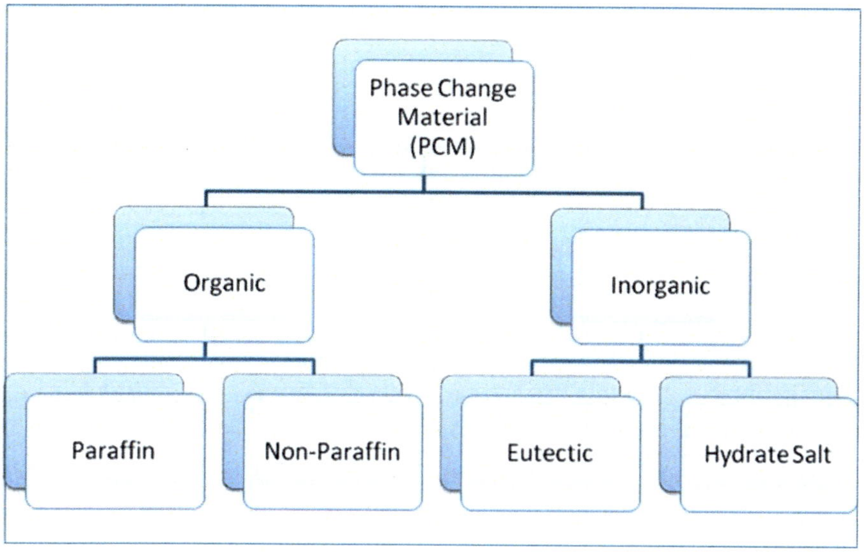

Fig. 1 Classification of phase change material

1.2.2 Non-paraffin

The non-paraffin PCM includes material like ester, glycols, alcohols and fatty acids. These materials are not used at high temperature because it is flammable. The uses of these materials are limited due to high corrosives, toxicity and instability at high temperature [8].

1.2.3 Eutectics

A eutectic is an inorganic PCM which is a composition of two or more material. It is achieved when the composition are solidified at a minimum freezing temperature. Some common eutectic mixture used as PCM is shown in Table 1.

Table 1 Eutectic mixture used as PCM [10]

Material	Latent heat (KJ/KgK)	Melting point (°C)
27%$LiNO_3$ + 68%NH_4NO_3 + 5%NH_4Cl	108	81.6
66.6%NH2CONH2 + 33.4%NH4Br	151	76
50%CaCl2 + 50%MgCl2·6H2O	95	25
34%$C_{14}H_{28}O_2$ + 66%$C_{10}H_2O_2$	147.7	24
53%NH_2CONH_2 + 47%NH_4NO_3	95	46
45%CaCl2·6H2O + 55%$CaBr_2$·6H2O	140	14.7

1.2.4 Hydrate Salt

Hydrate salt is a mixture of water and salt. These material have high-thermal conductivity (generally two times of paraffin), high heat of fusion per unit volume, slightly toxic and non-corrosive [8]. Some common hydrate salt used as PCM are $Mg(NO_3)_2 \cdot 6H_2O$, $NaS_2O_3 \cdot 5H_2O$, $CaCl_2 \cdot 6H_2O$, $KF \cdot 3H_2O$, etc.

2 Experimental Procedure

Heat energy storage tank is developed which consists of a cylindrical shell, heat pipes and solar receiver disk. The material chosen for the cylindrical shell and heat pipe were stainless steel grade 304 and copper, respectively. The specification of the cylindrical shell and heat pipes are mentioned in Tables 2 and 3, respectively. The heat pipes are connected to the solar receiver head and it is placed at the center of the cylindrical shell. After heating of receiver head for 4 h, the mixture of $NaNO_3$ (60%) and KNO_3 (40%) is completely melt in temperature range 220 °C. This melted mixture is works as heat transfer fluid, and start flowing through central heat pipe under pressure by capillary action. Central heat pipe is open at heater head. When storage tank is combined with the stirling engine, charging of PCM and energy supply to the Stirling engine takes place simultaneously. Extender tubes are used to improve the heat transfer rate from heat transfer fluid (HTF) to PCM. During charging process HTF flowing through the extender tube by capillary action and transfer heat to storage material. The selection of above PCM is based on the reason that it can work on both sensible and latent heat storage tank. The thermal storage is kept in open atmosphere and solar radiation heated the receiver head of the storage tank. The heat energy of sun is stored by the phase change material which is then used to run the Stirling engine connected at the other end of the storage tank. A k-type thermocouple is placed at the both end of the storage tank to measure the temperature (Fig. 2).

Table 2 Specifications of cylindrical shell [12]

Properties	Specifications
Material	Stainless steel 304 grade
Length of container	25 cm
Thickness	5 mm
Diameter	12 cm
Melting point	1450 °C
Thermal conductivity	16.2 W/mK
Thermal expansion coefficient	$17.2 * 10^{-6}$/K

Table 3 Specification of heat pipe [7]

Properties	Specifications
Material	Copper
No of pipe	1 at center and 8 outside the central pipe
Diameter of central pipe	2.5 cm
Length of central pipe	25.5 cm
Diameter of external pipe	1 cm
Length of external pipe	24 cm
Thickness	1 mm
Thermal conductivity	350 W/mK
Melting point	1083 °C
Specific heat	381 kJ/KgK

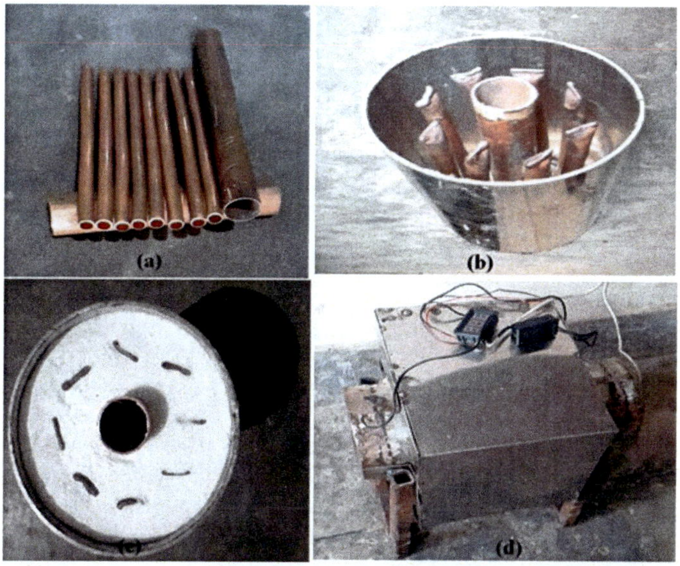

Fig. 2 **a** Heat pipes, **b** heat pipes placed into the cylindrical shell, **c** storage tank filled with PCM and **d** thermal energy storage tank with insulation

2.1 *Material Used*

In this experiment, two types of phase change material are used, first one is works as heat storage medium and second one is as heat transfer fluid which transfers heat from the receiver head to the heat storage medium.

2.1.1 Heat Storage Material

Eutectic mixture of NaCl–NaF is used as heat storage material for storing the heat energy. The eutectic blend is used in composition of 73% NaCl and 27% NaF on mass basis. The reason behind the use of eutectic blend of NaCl–NaF is that it can work as sensible and latent heat storage both. Due to this the size of storage tank gets reduce for storing same amount of heat using other PCMs. The properties of heat storage material are shown in Table 4.

2.1.2 Heat Transfer Fluid

Eutectic mixture of $NaNO_3$ (60%) and KNO_3 (40%) is used as heat transfer fluid (HTF). The properties of HTF are mentioned in Table 5.

3 Simulation Model

Ansys simulation is performed on the heat energy storage tank to analyze the temperature distribution within the tank. A number of assumptions were considered for the study is as follows:

- All the thermo-physical properties of the tank material are constant with respect to temperature.
- The effect of environmental condition is negligible.
- Constant heat flux is given to the receiver head.
- Temperature dependent properties of PCM are assumed as constant.

Since continuous heat is received by the receiver head so the eutectic mixture starts melting and molten material behaves like heat transfer fluid. The fluid flow in heat pipe is considered as laminar flow and its thermo-physical properties are considered linear and constant. The heat transfer in the heat pipe is modeled assuming finite thermal resistance offered by stainless steel. Inside the storage tank heat transfer is due to both conduction and convection. To minimize the heat loss the storage tank is insulated. The volumetric expansion of PCM inside the tank is neglected. The governing equations used for the above analysis are written as [7]:

Table 4 Properties of storage material [11]

Storage material	Eutectic mixture of NaCl–NaF
% composition	73% NaCl and 27% NaF (on mass basis)
Melting temperature	680 °C
Enthalpy	702 kJ/Kg

Table 5 properties of heat transfer fluid [12]

Material	Eutectic mixture of 60% $NaNO_3$ and 40% KNO_3. (on mass basis)
Melting temperature	220 °C
Boiling temperature	565 °C
Enthalpy	108.679 kJ/kg
Specific Heat capacity C_p	1.520 kJ/kg-K
Thermal conductivity of heat transfer fluid	0.53 W/mK
Density of heat transfer fluid	1804 kg/m^3

Continuity equation:

$$\frac{\partial \rho}{\partial t} + \nabla \rho V = 0 \tag{1}$$

Momentum equation:

$$\frac{\partial \rho V}{\partial t} + \nabla \rho v \times v = -\nabla p + \nabla \tau \tag{2}$$

Energy equation:

$$\frac{\partial \rho Cp(T)T}{\partial t} + \nabla \rho Cp(T)Tv = -(k(T)\nabla T) \tag{3}$$

The simulation modeling is performed on Ansys workbench R19.2 consists of geometry design, meshing and applying boundary conditions. The geometry of different part of storage tank is shown in Fig. 3. The accuracy of the simulation result largely depends upon mesh size irrespective of higher mesh size increased the processing time. A tetrahedral shaped mesh is used in this simulation. The meshing of the storage tank is shown in Fig. 4. The total number of nodes and elements of meshing are 385395 and 81356, respectively. For the better result of simulation finer mesh is employed.

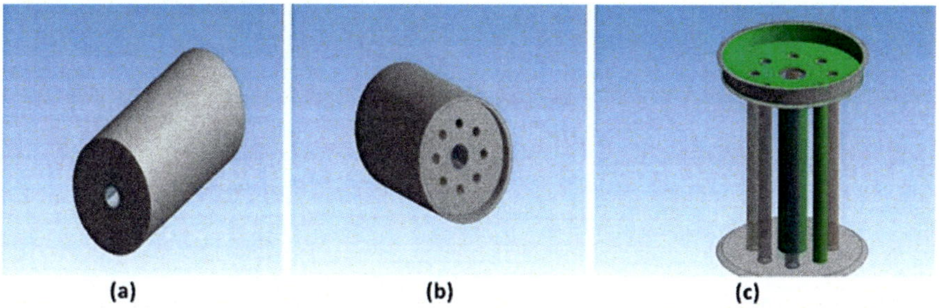

Fig. 3 Geometry of **a** cylindrical tank (Receiver side), **b** heater head side and **c** heat pipe arrangements

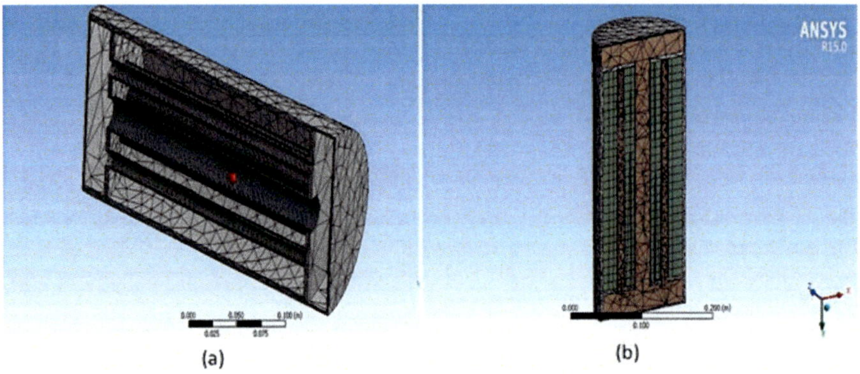

Fig. 4 Meshing of **a** tank and heat pipe **b** storage tank filled with PCM

4 Result and Discussion

The heat storage tank was heated with the solar radiation, concentrating the solar beam on the storage tank using Freshnel lens. It was seen that after heating the storage tank for 4 h, the eutectic mixture of phase change material ($NaNO_3$ and KNO_3) started melting at 220 °C and flow within the heat pipe carrying heat with itself. The k-type thermocouple is placed at the receiver head and heater head to measure the temperature of the receiver and heater head, respectively. The temperature measured was found to increase with the time as shown in Fig. 5.

The Ansys simulation result suggests that the temperature of heat storage tank increases uniformly, measured at an interval of 2 h which is shown in Figs. 6 and 7. The temperature of tank slowly increases and attains maximum value at 10 h. The uniform distribution of heat is achieved by the inclusion of heat pipes inside the storage tank. The molten phase change material flow through the heat pipe

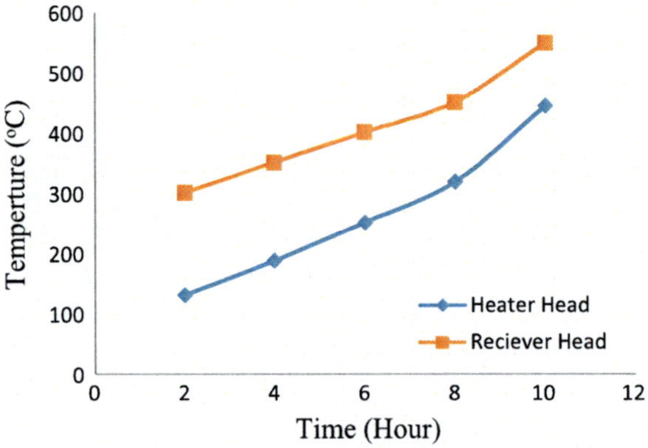

Fig. 5 Temperature measured of storage tank head at different time intervals

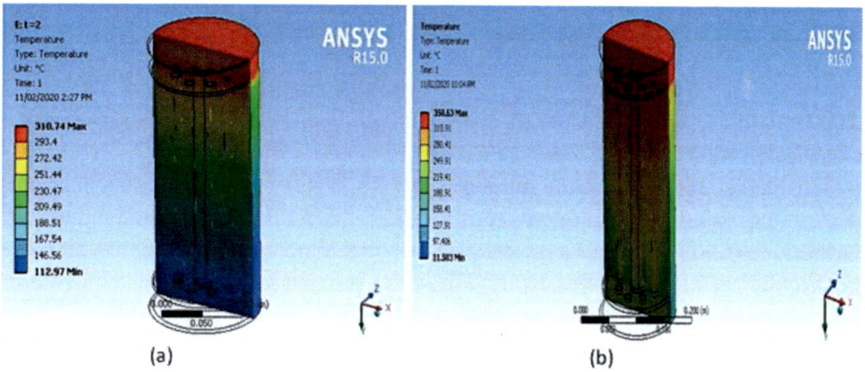

Fig. 6 Temperature distribution in the storage tank **a** after 2 h **b** after 4 h

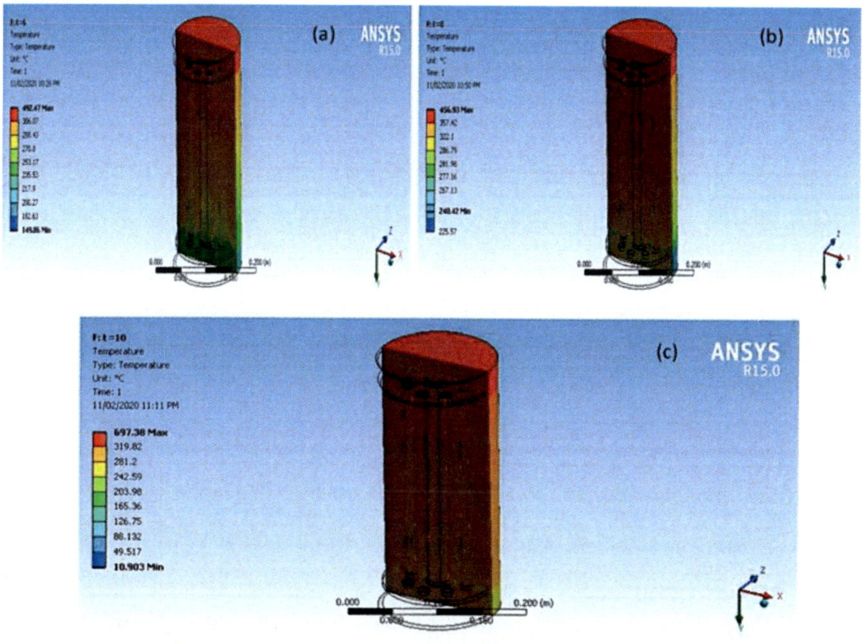

Fig. 7 Temperature distribution in the storage tank **a** after 6 h **b** after 8 h **c** after 10 h

distributing uniform heat in the tank. The objective of the current simulation work is to validate the experimental result collected during the heat transfer by the PCM in the storage tank. The three-dimensional model was validated by comparing the temperature of experimental and simulated work. The result showed that bulk temperature of experimental and simulated work deviates with maximum temperature of 135 °C.

5 Conclusion

This prototype thermal energy storage system shows temperature above 390 °C on its heater side which is capable to operate 1 KW gamma type Stirling engine for power generation. In this work, eutectic mixture of NaCl–NaF is used as heat storage media and eutectic mixture of $NaNO_3$–KNO_3 as a HTF. The charging scenario was investigated, after 10 h of heating; it is found that the temperature on heater side is 390 °C. Ansys simulation shows that temperature variation within the storage tank at an interval of every 2 h and it can be seen that the temperature distribution is uniform due to the installation of heat pipes. As the number of heat pipes increases heat transfer increases in the storage tank and charging process becomes smooth. The prototype is still need to improve the concept of design used in this system, so that the heat pipes are capable for maintaining the lowest possible temperature slope between PCM at top of storage tank and PCM at bottom of storage tank.

References

1. United States, Dep. Energy, Energy Inf. Adm.: International Energy Outlook 2010, pp 123–125, Washington, Julio (2010)
2. Medrano, M., Gil, A., Martorell, I., Potau, X., Cabeza, L.F.: State of the art on high-temperature thermal energy storage for power generation. Part 2—Case Studies **14**, 56–72 (2010)
3. Rathod, A.C., Bandela, P.C.V., Rehman, P.A.R.: Experimental Study on Phase Change Material based Thermal Energy Storage System, pp 2–7 (2017)
4. Sarbu, I., Sebarchievici, C.: Sustainability, no. February (2018)
5. Kenisarin, M.M.: Author's personal copy High-temperature phase change materials for thermal energy storage
6. Zhang, H., Baeyens, J., Cáceres, G., Degrève, J., Lv, Y.: Thermal energy storage : recent developments and practical aspects **53**, 1–40 (2016)
7. Bergman, T.L., Andraka, C.E.: IMECE2013-65487, pp 1–13 (2017)
8. Ricerca, D.D.I.: Modelling, design and analysis of innovative thermal energy storage systems using PCM for industrial processes, heat and power (2015)
9. Zhou, D., Zhao, C.Y., Tian, Y.: Review on thermal energy storage with phase change materials (PCMs) in building applications **92**, 593–605 (2012)
10. Cárdenas, B., León, N.: Latent heat based high temperature solar thermal energy storage for power generation. Energy Procedia **57**, 580–589 (2014)
11. Thermal Energy Storage for a Dish Stirling Concentrated no. July 2013 (2015)
12. Albanna, M.: Design of Heat Exchanger for Thermal Energy Storage Using High Temperature Phase Change Material no. April 2017 (2018)

Free Convective Heat Transfer with Boundary Slip Flow in a Nanofluid Along a Stretching Cylinder

Padam Singh, Vineet Kumar Sharma, and Manoj Kumar

1 Introduction

A study of heat transfer flow over an isothermal longitudinal circular cylinder moving in an incompressible and viscous fluid at rest has been done by Pop et al. [1]. It has been found that fluid suction/injection can affect the flow and heat transfer characteristics considerable. Later, the radial heat conduction along with heat generation in a vertical cylinder analyzed by Jilani et al. [2]. It was found that the radial temperature distribution in the boundary layer as well as the cylinder were remarkable significant especially with the inclusion of internal heat generation in the cylinder. Rani and Kim [3] analyzed the influence of temperature-dependent viscosity and Prandtl number on unsteady laminar free convection flow over a vertical cylinder. The heat transfer characteristics against the viscosity-variation parameter were analyzed with the help of average skin friction and Nusselt number and shown graphically. Bachok and Ishak [4] investigated the steady laminar flow caused by a stretching cylinder immersed in an incompressible viscous fluid with prescribed surface heat flux. It has been observed that the surface shear stress and the heat transfer rate at the surface increase as the curvature parameter increases. Steady laminar magnetohydrodynamic forced convective heat and mass transfer about a horizontal slender cylinder studied by Kaya [5]. It has been determined that the local skin friction coefficient, the local heat transfer coefficient, and the local mass transfer coefficient increase with an increase in the magnetic parameter and transverse curvature parameter. Kabeir [6]

P. Singh
Department of Applied Sciences, Galgotias College of Engineering and Technology, Greater Noida, India

V. K. Sharma (✉)
Department of Applied Sciences, G. L. Bajaj Institute of Technology and Management, Greater Noida, India

M. Kumar
G. B. Pant University of Agriculture and Technology, Pantnagar, Uttrakhand, India

L. M. Das et al. (eds.), *Recent Trends in Thermal Engineering*, Lecture Notes in Mechanical Engineering,
https://doi.org/10.1007/978-981-16-3428-4_10

studied the diffusion-thermo and thermal diffusion effects on heat and mass transfer by boundary layer flow over a stretching cylinder embedded in a porous medium in the presence of chemical reaction effect. It was found that the local Nusselt number decreased, whereas Sherwood increased due to increases in either the Reynolds number or Darcy number. Also, increases in the values of Soret number produced increases in the local Nusselt number and decreases in the local Sherwood number. Chamaka [7] studied steady, laminar, incompressible, hydromagnetic boundary layer flow, and heat and mass heat transfer over a permeable cylinder moving or stretched with a linear velocity in the presence of heat generation/absorption, chemical reaction, suction/injection effects, and uniform transverse magnetic field. It was found that the skin friction coefficient was positive for velocity ratios greater than unity and negative for velocity ratios less than unity. When the velocity ratio is less than unity, the skin friction coefficient increased as either of velocity ratio or the curvature parameter increased while it decreased as the Hartmann number or the suction/injection coefficient increased. Elbashbeshy et al. [8] analyzed effect of magnetic field on flow and heat transfer of an incompressible viscous fluid over a stretching horizontal cylinder in the presence of heat source or sink with suction/injection. It was found that the skin friction can be minimized by decreasing the magnetic field intensity and increasing the radius of the stretching cylinder and the suction/injection velocity. The effects of thermo-physical properties on the axisymmetric flow of a viscous fluid induced by a stretching cylinder in the presence of internal heat generation/absorption studied by Vajravelu et al. [9]. The flow and heat transfer characteristics reveal many interesting behaviors that warrant further study on the axisymmetric flow phenomena.

Keeping the above curious findings, the present work analyzes the influence of slip parameter and local free convection parameter on heat transfer flow in alumina water nanofluid along a stretching cylinder.

2 Mathematical Description

The physical model is drawn here along with flow configuration and coordinate system. Assuming that the fluid is infinite in extent of positive x-direction. Temperature and concentration are denoted by T and C with suffix 'w', '∞' for cylinder surface and free stream, respectively, and u and v denotes velocity components in the direction of x and r, respectively. The governing equations take the following form (Fig. 1):

$$\frac{\partial}{\partial x}\left(\rho_{nf} r u\right) + \frac{\partial}{\partial r}\left(\rho_{nf} r v\right) = 0 \tag{1}$$

$$u\frac{\partial u}{\partial x} + v\frac{\partial u}{\partial r} = \frac{\upsilon_{nf}}{r}\frac{\partial}{\partial r}\left(r\frac{\partial u}{\partial r}\right) + g(\beta_T)_{nf}(T - T_\infty) \tag{2}$$

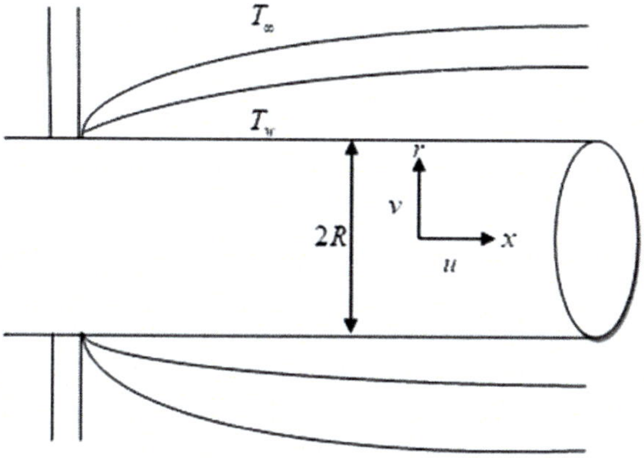

Fig. 1 Flow configuration with coordinate system

$$u\frac{\partial T}{\partial x} + v\frac{\partial T}{\partial r} = \frac{k_{nf}}{r\rho_{nf}(C_p)_{nf}}\frac{\partial}{\partial r}\left(r\frac{\partial T}{\partial r}\right) \tag{3}$$

With the boundary conditions:

$$u = U_0\frac{x}{L} + B_0\upsilon_{nf}\frac{\partial u}{\partial r};\ v = 0, T = T_w \text{ at } r = R \text{ and } u \to u_\infty;\ T \to T_\infty \text{ as } r \to \infty \tag{4}$$

where $T_w = T_\infty + T_0\left(\frac{x}{L}\right)^n$ is surface temperature, U_0 and T_0 denotes the reference velocity and temperature, respectively, $(\beta_T)_{nf}$ is thermal expansion coeficient of the nanofluid, g is gravitational acceleration, n is temperature exponent (Table 1).

Suffix 'f' and 's' denotes the pure water and solid alumina, respectively, and ϕ is volume fraction of the nanofluid (Table 2).

Table 1 Nanofluid properties

Dynamic viscosity [10]	$\mu_{nf} = \mu_f/(1-\phi)^{2.5}$
Density [11]	$\rho_{nf} = (1-\phi)\rho_f + \phi\rho_s$
Specific heat capacity [12]	$(C_p)_{nf} = (1-\phi)(C_p)_f + \phi(C_p)_s$
Kinematic viscosity [13]	$\upsilon_{nf} = \mu_{nf}/\rho_{nf}$
Thermal diffusivity	$\alpha_{nf} = k_{nf}/\rho_{nf}(C_p)_{nf}$
Thermal conductivity [14]	$(k_s/k_f) + 2 - \phi(1-(k_s/k_f))k_{nf} = k_f(k_s/k_f) + 2 - 2\phi(1-(k_s/k_f))$

Table 2 Physical properties of alumina and water at 20 °C are given below

	Density (Kg/m^3)	Thermal conductivity (W/m/K)	Specific heat (J/Kg/K)
Alumina	3970	36	769
Water	1000.52	0.597	4181.8

3 Problem Solution

Converting the governing Eqs. (1)–(3) with boundary conditions (4) into non-dimensional form by introducing the following similarity transformation:

$$\eta = \frac{r^2 - R^2}{2R}\left(\frac{U}{x\upsilon_{nf}}\right)^{1/2}, \psi = Rf(\eta)\left(x\upsilon_{nf}U\right)^{1/2}, \\ \frac{T - T_\infty}{T_w - T_\infty} = \theta(\eta), u = \frac{1}{r}\frac{\partial\psi}{\partial r} \text{ and } v = -\frac{1}{r}\frac{\partial\psi}{\partial x} \tag{5}$$

After using above transformation, Eqs. (1)–(3) reduced to the nonlinear differential equations as follows:

$$(2K\eta + 1)f'''(\eta) + (f(\eta) + 2K)f''(\eta) - f'^2 + g_t\theta(\eta) = 0 \tag{6}$$

$$(2K\eta + 1)\theta''(\eta) + 2K\theta' + \Pr(f\theta' - nf'\theta) = 0 \tag{7}$$

and the boundary conditions (4) reduced as follows:

$$f(0) = 0, f'(0) = 1 + \gamma f''(0), \theta(0) = 1 \text{ and } f'(\infty) = 0, \theta(\infty) = 0 \tag{8}$$

Here $K = \frac{1}{R}\sqrt{\frac{L\upsilon_{nf}}{U_0 R^2}}$ is the curvature parameter, $g_t = \frac{g(\beta_T)_{nf}(T_w - T_\infty)}{\mathrm{Re}\upsilon_{nf}}$ is local free convection parameter, $\Pr = \frac{\upsilon_{nf}\rho_{nf}(C_p)_{nf}}{k_{nf}}$ is the Prandtl number, $\gamma = B_0\sqrt{\frac{\upsilon_{nf}U_0}{L}}$ is slip parameter with reference value B_0, $\mathrm{Re} = \frac{U_0^2 x}{\upsilon_{nf}}$ is local reynolds number.

To solve the nonlinear differential Eqs. (6)–(7) subject to the boundary conditions (8), adaptive Runge-Kutta method with shooting technique has been applied. The domain of the problem is discretized, and the boundary conditions for $\eta = \infty$ are replaced by $\eta_{max} = 10$ a sufficiently large value of η at which the boundary conditions (8) are satisfied. We first convert the nonlinear Eqs. (6)–(7) into first order ordinary linear differential equations. There are three conditions on the boundary $\eta = 0$ and two conditions at $\eta \to \infty$. To get the solution of the problem, two more initial conditions have been found by shooting technique. The value of unknowns $f''(0)$, and $\theta'(0)$ at which solution agrees with the boundary conditions (8) has been given in Table 4. Adaptive Runge-Kutta method estimates the truncation error at each integration step and automatically adjusts the step size to keep the error within limit of tolerance. The formula used to measure the actual conservative error is defined by $e(h) = \sqrt{\frac{1}{5}\sum_{i=1}^{5} E_i^2(h)}$ here $E_i(h)$ is the measure of error in the dependent variable y_i. Per-step error control is achieved by adjusting the increment h by $h_{i+1} = 0.9h_i(\varepsilon/e(h_i))^{1/5}$; $i = 1, 2, 3, 4, 5$. so that it is approximately equal to the desired tolerance $\varepsilon = 10^{-6}$. The assumed initial step size is $h = 0.01$.

4 Results and Discussion

The computations have been made for velocity, shear stress, temperature, and temperature gradient profiles corresponding to various values of slip parameter (γ) and free convection parameter (g_t). The physical and thermal properties of alumina water nanofluid corresponding to different volume fraction have also been given in Table 3. The values of $f''(0)$ and $\theta'(0)$ obtained by shooting have been tabulated in Table 4. The present study deals with alumina water nanofluid of Pr = 6.6556406 corresponds to 0.05 volume fraction (ϕ) at 20 °C.

For the verification of accuracy of the numerical scheme, a comparison has been made corresponding to the heat transfer coefficient $[-\theta'(0)]$ for several values of temperature exponent (n) at $K = 0$ (i.e., for a flat plate) and $\gamma = 0$ (i.e., in case of no boundary layer slip) with the previous published results of Ali [15], Ishak and Nazar [16], Mukhopadhyay and Gorla [17] which has been found in good agreement as given in Table 5.

Figures 2, 3, 4 exhibit the velocity, shear stress, temperature, and temperature gradient profiles, respectively, for various values of slip parameter (i.e.,$\gamma = 0, 0.5, 1.0, 1.5$). It is noticed that in Fig. 2, the rate of transport decreases with

Table 3 Physical parameters for Al_2O_3 nanofluid at 20 °C

Vol. Fr. ϕ	Thermal conductivity K_{nf}	Dynamic viscosity μ_{nf}	Prandtal number Pr	Thermal diffusivity α_{nf}	Kinematic viscosity ϑ_{nf}	Density ρ_{nf}	Heat capacity C_{pnf}
0.00	0.5970000	1.0020000	7.0186994	0.1426872	1.0014792	1000.5200	4181.800
0.01	0.6142114	1.0274950	6.9385106	0.1437426	0.9973600	1030.2148	4147.672
0.02	0.6317568	1.0539075	6.8622847	0.1448988	0.9943372	1059.9096	4113.544
0.03	0.6496460	1.0812805	6.7898403	0.1461537	0.9923606	1089.6044	4079.416
0.04	0.6678893	1.1096592	6.7210103	0.1475057	0.9913875	1119.2992	4045.288
0.05	0.6864972	1.1390917	6.6556406	0.1489536	0.9913818	1148.9940	4011.160
0.06	0.7054809	1.1696288	6.5935894	0.1504967	0.9923135	1178.6888	3977.032
0.07	0.7248517	1.2013245	6.5347257	0.1521346	0.9941582	1208.3836	3942.904
0.08	0.7446217	1.2342358	6.4789292	0.1538674	0.9968963	1238.0784	3908.776
0.09	0.7648033	1.2684233	6.4260890	0.1556954	1.0014792	1000.5200	4181.800
0.10	0.7854095	1.3039515	6.3761029	0.1576193	0.9973600	1030.2148	4147.672

Table 4 Values of $f''(0)$ and $-\theta'(0)$ for $K = 0.2$, $n = 1$ and $\Pr = 6.6556406$ at $\phi = 0.05$

Slip parameter (γ)	Free convection parameter (g_t)	$f'(0)$	$-\theta'(0)$
0	0.2	−0.47090	1.3767
0.5	0.2	−0.34990	1.2549
1.0	0.2	−0.28150	1.1749
1.5	0.2	−0.23660	1.1159
0.2	0	−0.43009	1.3148
0.2	0.5	−0.38647	1.3291
0.2	1.0	−0.34412	1.3434
0.2	2.0	−0.26290	1.3697

Table 5 Values of $-\theta'(0)$ for several values of temperature exponent (n) in case of flat plate (i.e., $K = 0$) and without slip (i.e., $B = 0$)

n	Ali [15]	Ishak and Nazar [16]	Mukhopadhyay and Gorla [17]	Present results
0	0.5801	0.5820	0.5821	0.5820
1	0.9961	1.0000	1.0000	1.0000
2	1.3269	1.3333	1.3332	1.3334

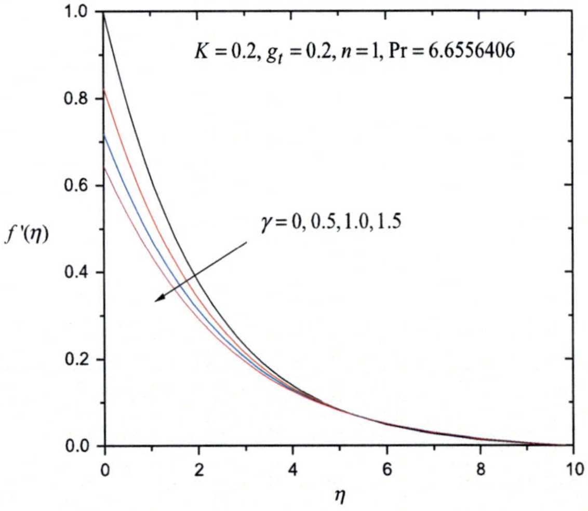

Fig. 2 Effect of slip parameter on velocity profile

increasing distance (η), and the velocity vanishes at some maximum distance ($\eta_{\max} = 10$). The velocity decreases with increasing value of slip parameter. Shear stress is larger at the surface of stretching cylinder for a larger value of slip parameter. Shear stress increases with an increase in the slip parameter which produces a larger skin friction coefficient as shown in Fig. 3. Temperature gradient increases up to $\eta = 0.7786$ and then decreases asymptotically with an increase in

Fig. 3 Effect of slip parameter on shear stress

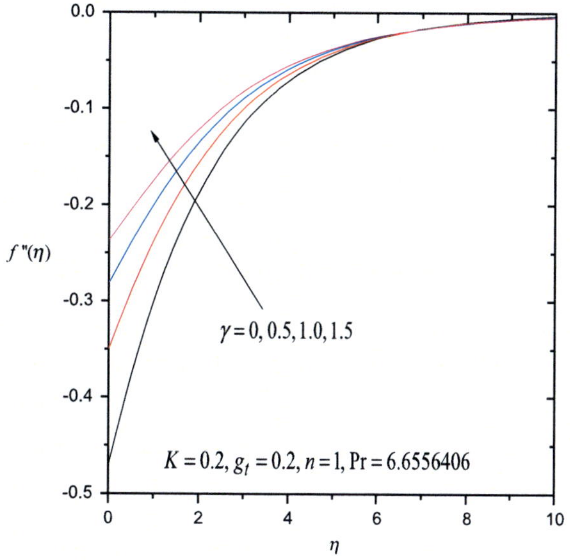

Fig. 4 Effect of slip parameter on temperature gradient

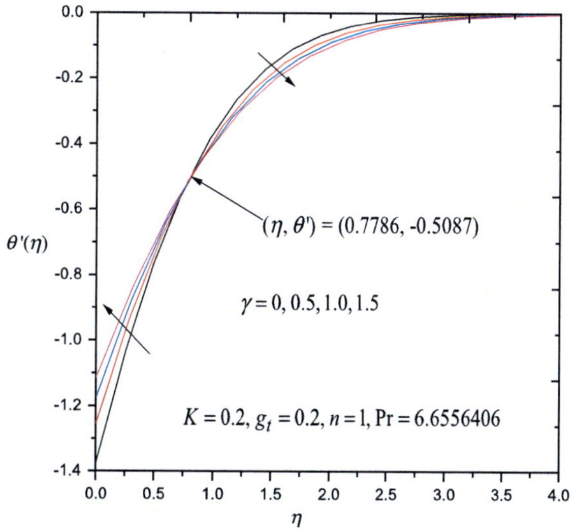

the slip parameter, as depicted in Fig. 4. The temperature gradient is maximum at the surface for larger value of slip parameter which produces a maximum heat transfer rate at the surface and larger value of Nusselt number(Nu_x) since $Nu_x \mathrm{Re}_x^{1\backslash 2} = -\theta'(0)$.

In Fig. 5, the temperature is found to increase with increasing value of slip parameter. Effect of free convection parameter (g_t) on velocity is depicted in Fig. 6, and it is noticed that the velocity increases with increasing value of free convection

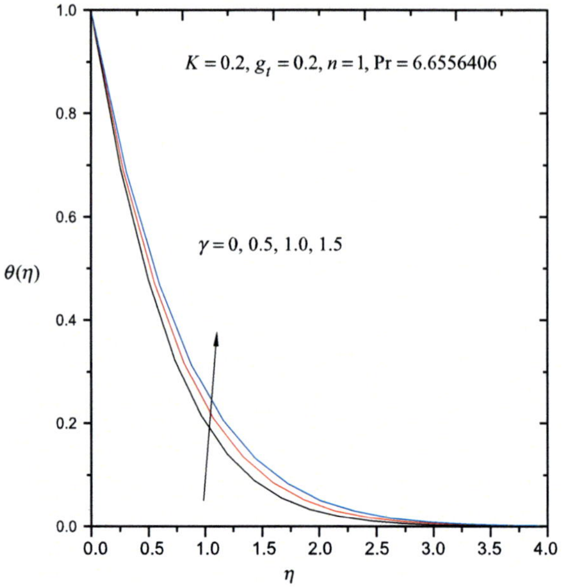

Fig. 5 Effect of slip parameter on temperature

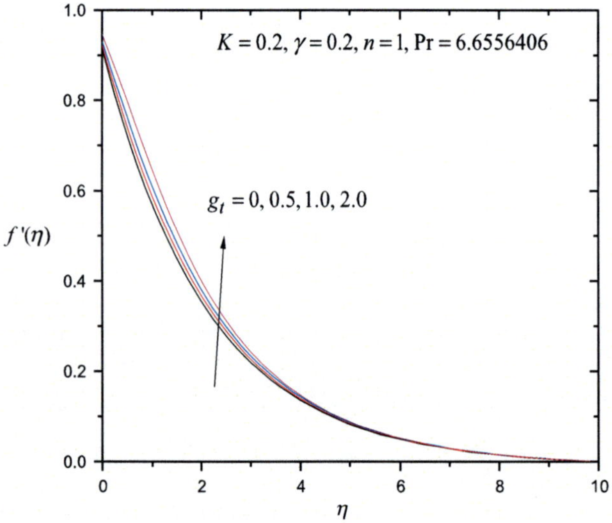

Fig. 6 Effect of free convection parameter on velocity

parameter. Figure 7 shows the effect of free convection parameter on shear stress. It has been observed that the shear stress decreases up to $\eta = 0.6888$ from the cylinder surface and then decreases asymptotically. It has been observed that the

Fig. 7 Effect of free convection parameter shear stress

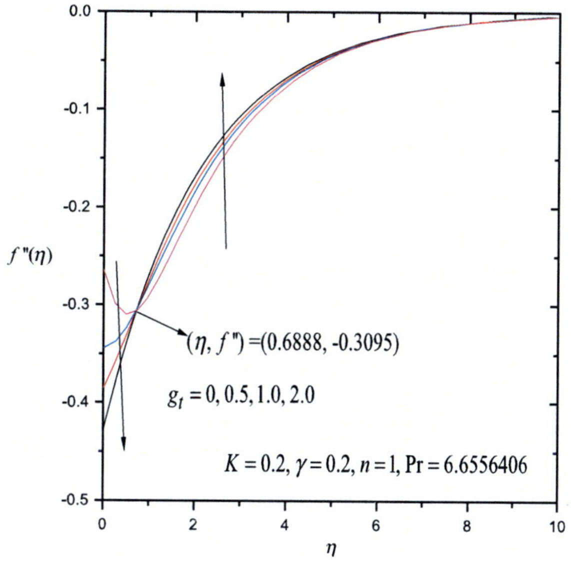

Fig. 8 Effect of free convection parameter on temperature

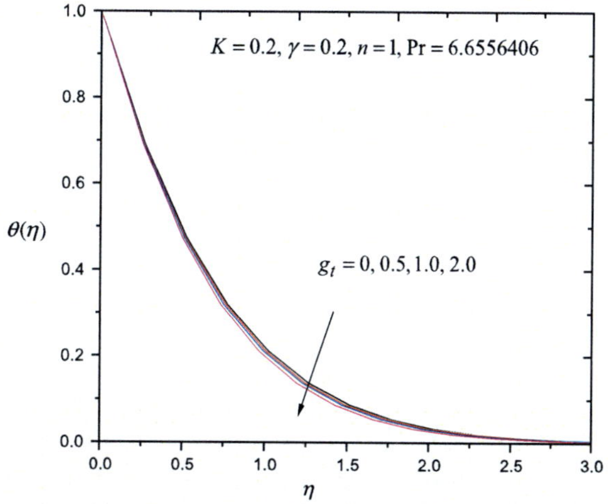

temperature decreases with increasing free convection parameter as shown in Fig. 8. In Fig. 9, temperature gradient decreases up to $\eta = 0.7893$ and then increases with an increase in free convection parameter.

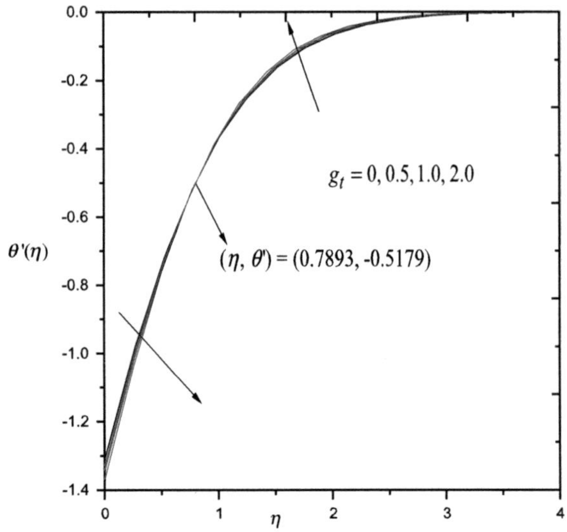

Fig. 9 Effect of free convection parameter on temperature gradient

5 Conclusion

Heat transfer in an electrically conducting and viscous alumina water nanofluid over a stretching cylinder under slip conditions is considered. The governing nonlinear partial differential equations have been transformed into a system of ordinary differential equations by using similarity transformation and then solved numerically using the fifth order Runge–Kutta method along with the shooting technique. From the numerical computation of effect of slip parameter and free convection parameter, it is concluded that.

- The comparison made corresponding to the heat transfer coefficient for several values of temperature exponent with the previous published work found in good agreement.
- The velocity decreases with increasing value of slip parameter, and the shear stress is larger at the surface of stretching cylinder for a larger value of slip parameter. However, shear stress increases with an increase in the slip parameter which produces a larger skin friction coefficient.
- Temperature gradient increases up to a fixed range and then decreases asymptotically with an increase in the slip parameter. The temperature gradient is higher at the surface for larger value of slip parameter which produces a maximum heat transfer rate at the surface and larger value of Nusselt number. The velocity increases with increasing value of free convection parameter and the temperature decreases with increasing free convection parameter.

References

1. Pop, T., Watanabe, H.: Taniguchi, laminar boundary layer flow and heat transfer along a moving cylinder with suction or injection. Technische Mechanik **15**(2), 99–106 (1995)
2. Jilani, G., Jayaraj, S., Adeel Ahmad, M.: Conjugate forced convection-conduction heat transfer analysis of a heat generating vertical cylinder. Int. J. Heat Mass. Transfer. **45**, 331–341 (2002)
3. Rani, H.P., Kim, C.N.: Transient free convection flo over an isothermal vertical cylinder with temperature dependent viscosity. Korean J. Chem. Eng. **25**(1), 34–40 (2008)
4. Bachok, N., Ishak, A.: Flow and heat transfer over a stretching cylinder with prescribed surface heat flux. Malays. J. Math. Sci. **4**(2), 159–169 (2010)
5. Kaya, A.: Heat and mass transfer from a horizontal slender cylinder with a magnetic field effect. J. Therm. Sco. Tech. **31**(2), 73–78 (2011)
6. El-Kabeir, S.M.M.: Soret and dufour effects on heat and mass transfer due to a stretching cylinder saturated porous medium with chemically-reactive species. Latin Am. Appl. Res. **41**, 331–337 (2011)
7. Chamkha, A.J.: Heat and mass transfer from MHD flow over a moving permeable cylinder with heat generation or absorption and chemical reaction. Commun. Numer. Anal. **2011** (cna-00109), 1–20 (2011)
8. Elbashbeshy, E.M.A., Emam, T.G., El-Azab, M.S., Abdelgaber, K.M.: Effect of magnetic field on flow and heat transfer over a stretching horizontal cylinder in the presence of a heat source/sink with suction/injection. J. Appl. Mech. Engg. **1**(1), 1–5 (2012)
9. Vajravelu, K., Prasad, K.V., Santhi, S.R., Umesh, V.: Fluid flow and heat transfer over a permeable stretching cylinder. J. Appl. Fluid Mech. **7**(1), 111–120 (2014)
10. Einstein, Investigation on the Theory of the Brownian Movement. Dover, New York (1956)
11. Drew, D.A., Passman, S.L.: Theory of multicomponent fluids. Springer (1999)
12. Xuan, Y., Roetzel, W.: Conceptions for heat transfer correlation of nanofluids. Int. J. Heat Mass Transfer **43**, 3701–3707 (2000)
13. Mamut, E.: Characterization of heat and mass transfer properties of nanofluids. Rom. J. Phys. **51**, 5–12 (2006)
14. Patel, H.E., Sundararajan, T., Pradeep, T., Dasgupta, A., Dasgupta, N., Das, S.K.: A micro convection model for thermal conductivity of nanofluids. Pramana J. Phy. **65**, 863–869 (2005)
15. Ali, M.E.: Heat transfer characteristics of a continuous stretching surface. Heat Mass Transfer **29**, 227–234 (1994)
16. Ishak, A., Nazar, R.: Laminar boundary layer flow along a stretching cylinder. Euro. J. Sci. Resc. **36**(1), 22–29 (2009)
17. Mukhopadhyay, S., Gorla, R.S.R.: Slip effects on boundary layer flow and heat transfer along a stretching cylinder. Int. J. Appl. Mech. Engg. **18**(2), 447–459 (2013)

Cooling Methods for Solar Photovoltaic Modules Using Phase Change Materials: A Review

Amit Kumar and Lalta Prasad

1 Introduction

The hazardous effects of power generation on environment can be mitigated by reducing our dependence on fossil fuels. Power generation using renewable sources is gaining a lot of attention recently owing to fast depleting fossil fuel reserves and growing focus towards clean energy and sustainability. It is the need of time to exploit more and more renewable energy sources and reduce our dependence on fossil fuels for power generation to meet the future electricity demand while causing minimal damage to our environment [1]. Global renewable power generation was increased by 6.1% in 2018 as compared to 2017 [2]. In 2019, share of renewables in global electricity generation reached 10.4%, surpassing nuclear power for the first time [3].

Among all the renewable means of power generation such as solar, wind power, geothermal, biomass, and small-hydro, solar energy has been observed as a viable alternative for power generation with a large potential to achieve future goals of clean energy production [4]. According to an estimate, Earth receives a huge amount of energy as sunlight in one hour which is sufficient for the world for an entire year [5]. Hence, the potential of solar energy is infinite, and the only limit is our ability to convert solar radiation into more usable form. Figure 1 provides renewable power breakup in India which shows that nearly 40% of installed renewable power in India comes from solar energy conversion.

Employing solar photovoltaic panels for power generation presents several advantages over solar thermal method as they are silent, static, and directly provide high grade of energy. However, large initial cost and low efficiency of solar panels remain big barriers in widespread deployment of this technology. Table 1 shows

A. Kumar (✉) · L. Prasad
Department of Mechanical Engineering, National Institute of Technology, Srinagar (Garhwal), Uttarakhand 246174, India
e-mail: amitsharma.phd19@nituk.ac.in

L. M. Das et al. (eds.), *Recent Trends in Thermal Engineering*,
Lecture Notes in Mechanical Engineering,
https://doi.org/10.1007/978-981-16-3428-4_11

Fig. 1 **a** Renewable power breakup July 2020 [6] **b** Growth of PV installed capacity in India till March 31, 2020 [7]

Table 1 Efficiencies of various types of solar PV cells [10]

Sr. No	PV cell technology	Conversion efficiency at lab scale (%)
1	Carbon nano-tube (CNT)	10–12
2	Thin film	14–22
3	Crystalline Si	21–25
4	Multi-Junction cells	31–38
5	Perovskite solar cells	22
6	Dye-sensitized solar cells	11
7	Organic solar cells	10

efficiency of various types of PV cells at standard test conditions. While the efficiency of solar thermal power systems may go beyond 35% [8], the maximum conversion efficiency of most popular silicon solar panels is still restricted to 26% at standard test conditions [9]. The actual conversion efficiency achieved by solar PV panels is even lesser than the value attained at standard test conditions.

1.1 Effect of High Temperature on PV Modules

The efficiency of a solar photovoltaic module depends on several factors such as cell material and technology, radiation intensity, ambient temperature, sun tracking, shading, soiling of module, and equipment efficiency. Module surface temperature is a major factor which reduces the conversion efficiency especially at high solar intensity. At STC, the module is kept at 25 °C [9] while in actual practice, module temperature may go well beyond 50 °C during peak hours of solar insolation. PV module efficiency declines significantly with the rise in module temperature. Generally, a 0.5% decline in conversion efficiency occurs with 1 °C increase in

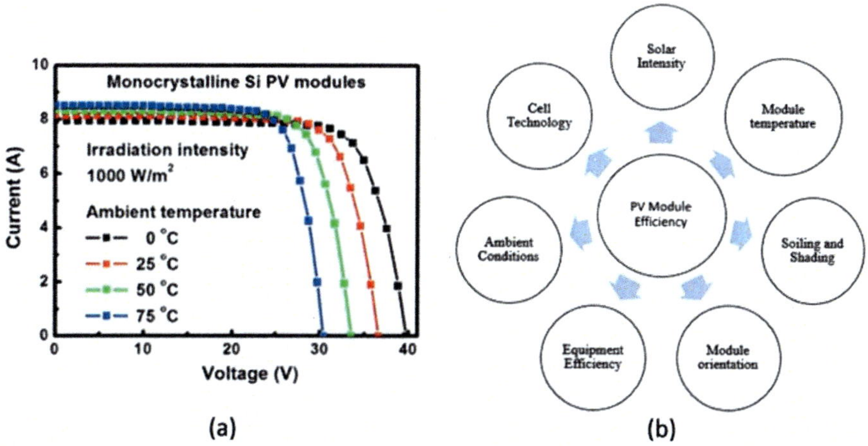

Fig. 2 **a** Effect of temperature on Si solar cell performance [13] **b** Factors affecting PV module efficiency

module temperature [11, 12]. It has also been observed that PV modules working at relatively low temperature works for longer period. So, for high conversion efficiency and longer life, PV module temperature should be kept low during operation. So, a thermal management system for PV modules is essential to ensure high performance and longer life. Figure 2 represents various factors which affect the PV module efficiency and the effect of module temperature on electrical characteristics of a typical Si module.

Several researchers have conducted experiments to quantify and predict the degradation of module performance due to high module temperature. Table 2 summarizes a few of those studies.

1.2 Types of PV Cooling Methods

Various methods of PV module cooling have been developed and implemented by several researchers. These methods can be broadly categorized as active methods and passive methods as represented in Fig. 3. Active cooling methods require external power for their operation because they generally use a coolant circulation over front and/or back side of modules to carry away the heat. They are generally termed as photovoltaic thermal (PV/T) systems. Several PV/T configurations with forced-air cooling, water cooling, water spray, and nanofluids circulation have been extensively reported in the literature. Passive cooling methods do not require external power. These include natural convection cooling, cooling fins, thermoelectric modules (TEM), phase change material (PCM) cooling, spectral beam splitting, etc. Passive cooling methods are less costly but also less effective when compared to active cooling.

Table 2 Studies on degradation of PV modules due to high temperature

Study type	Short description	Effect of module temperature	References
Experimental and numerical	PV module's output was estimated based on the relationship between environmental factors and basic meteorological data of local area	Operating temperature was found to be the major factor for performance degradation in PV modules	[14]
Review	Studied several mathematical correlations between temperature of a PV module and its conversion efficiency	Keeping the module temperature within permissible limit would ensure high conversion efficiency	[15]
Experimental	Studied the degradation of module performance due to various environmental factors, viz., solar intensity, ambient temperature, relative humidity, and soiling of PV module surface	Significant reduction in electrical output of PV module was observed due to high ambient temperature in single day of operation	[16]
Experimental	Studied the drop in PV module output and efficiency with rise in its temperature at various irradiance levels ranging from 400 W/m^2 to 1000 W/m^2	At 1000 W/m^2 irradiance level, the PV module was able to attain maximum temperature of 56 °C resulting in 3.13% reduction in its efficiency	[17]
Experimental	Investigated the effect of PV module surface temperature on its conversion efficiency	Module efficiency was reduced from 12.07% to 10.7% when module temperature rose from 14.9 °C to 51.3 °C	[18]

Selection of a cooling method for PV modules depends on several factors such as ambient conditions, available resources, system durability, and cost to benefit ratio. No clear standards have been reported for cooling of PV modules.

The present work attempts to review and summarize recent research articles focused on phase change materials based cooling systems for PV modules. Research articles on the use of more than one PCM, composite PCM, and PCM with active water cooling have been discussed. Moreover, various methods to enhance thermal conductivity of PCMs such as construction of internal fins, porous metal structures, and use of nanoparticles have also been discussed. Finally, few suggestions for future work on PV module cooling have been included .

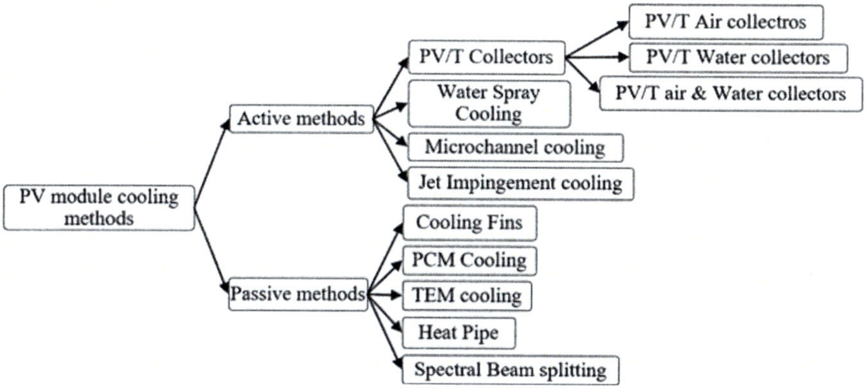

Fig. 3 Various cooling methods for solar PV modules [19]

2 Studies on Pure PCM Used with PV Modules

Phase change materials (PCMs) are most suitable for reducing the temperature of PV modules as they can be easily placed on the rear side of a module by constructing a suitable container. A general schematic of such a system has been shown in Fig. 4. Several designs of containers with different phase change materials have been developed by researchers for PV-PCM modules. Some noteworthy studies are summarized in Table 3.

Rajvikram et al. tried to augment PV module efficiency by placing a container on backside of the module filled with PCM-HS29 having melting point of 29 °C [32]. The container was further attached to an aluminum heat sink to dissipate the heat stored in PCM to the atmosphere. Mahamudul et al. [33] conducted numerical simulation on PV modules integrated with PCM for its thermal regulation for Malaysian weather conditions. Paraffin RT35 with 2 cm layer thickness was placed on rear side of the PV module. Simulation results were validated by comparing them with the experimental data. Maximum temperature reduction of 10 °C was obtained for 6 h of module operation.

The PV-PCM system was found to be effective in keeping low module temperature thus ensuring high output. Maximum efficiency of nearly 17% was achieved by employing the PV-PCM cooling system with aluminum heat sink. Hasan et al. developed a PV-PCM module for extremely hot climate of UAE [34]. It was built by placing a paraffin container on the module back. Melting temperature range of PCM was 38–43 °C which was optimum according to ambient conditions of the location. Tests were conducted for yearlong period in order to evaluate performance and economic feasibility of PV-PCM module. PV-PCM system showed variable performance during different months of the year. It was especially least effective in months of extreme weather owing to its partial melting and solidification. Maximum drop of 13 °C and average drop of 10 °C in module temperature were achieved resulting in 5.9% increase in annual power output.

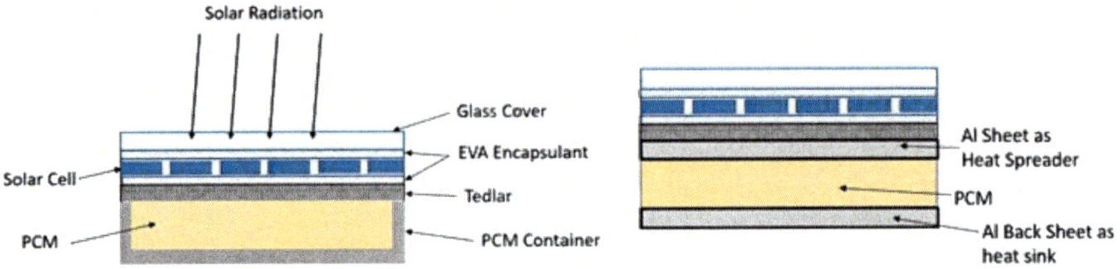

Fig. 4 Schematic representation of PCM container in PV-PCM systems

Table 3 Summary of noteworthy research articles on PCM based cooling systems for Solar PV modules

Author	Cooling method and PCM used	Major findings
Al-Waeli et al. [20]	Paraffin wax	6.5% improvement in electrical efficiency
Indartono et al. [21]	Petroleum jelly	Average increase of 7.3% and 6% in electrical output and conversion efficiency, respectively
Choubineh et al. [22]	Salt hydrate (PCM 32/280) with forced-air cooling	9% increase in PV electrical efficiency
Bayrak et al. [19]	$CaCl_2.6H_2O$ with fins and TEM	7.7% improvement in module electrical output
Waqas et al. [23]	Macro-encapsulated organic PCM	9 °C reduction in peak module temperature and 2% improvement in conversion efficiency
Hasan et al. [24]	RT 42 with water cooling	9 °C reduction in peak temperature and 1.3% improvement in conversion efficiency
Hossain et al. [25]	Lauric acid with water cooling	1.2% increase in PV module conversion efficiency
Sarafraz et al. [26]	Paraffin with multiwalled CNT	0.2% (by weight) carbon nanotube suspension in PCM showed maximum improvement in performance of PV-PCM module
Abdollahi and Rahimi [27]	Coconut oil and sunflower oil Boehmite nanoparticles	48.23% increase in module output
Al-Waeli et al. [28]	Paraffin wax and SiC nanoparticles	PV efficiency was improved when SiC nanoparticles were used as coolant
Eisapour et al. [29]	Ag/water nanofluid and microencapsulated nano-PCM slurry	8.58% and 0.6% increase in thermal efficiency and electrical efficiency, respectively, compared with simple water cooling
Sharma et al. [30]	CuO nanoparticles with RT 42 PCM in micro-finned heat sink $CaCl_2H_{12}O_6$ with Al_2O_3	18.5% reduction in PV module temperature
Salem et al [31]	Nanoparticle and/or water cooling	Water cooled PV module which utilized 1% mixture of nanoparticles in PCM showed the highest performance

Kawtharani et al. [35] experimentally investigated capric acid as PCM to regulate its temperature in indoor test conditions. The organic fatty acid-based PCM had melting point ranging from 27 to 32 °C. It was found to be effective for keeping module temperature within permissible limit. Further, a numerical simulation model was also developed and validated to theoretically predict the temperature variation of PV module and its electrical characteristics. The PV-PCM module was able to save 16% energy as compared to the conventional PV module. Hendricks and Van Sark [36] conducted tests on several PCMs in order to investigate the possibility of using them to limit the PV module temperature. A simple heat balance model was applied to PV-PCM modules to analyze net energy gain for different runs of cooling system. Several numerical simulations were run on PCMs namely: Rubitherm RT-27, RT-42 BASF micronal DS5001, and Thermusol HD35. Rubitherm PCMs were found to be effective in keeping relatively low module temperature. However, economic feasibility was not reached for a practical scenario. Integration of PCMs in BIPV systems for inside climate control of the building was suggested as future scope for further research. Hasan et al. designed and developed PV-PCM systems with two different PCMs and conducted experimental investigations on these systems in cold as well as hot climatic conditions [37]. PCM stores the excessive heat generated by the module for sufficiently long duration. While the PV-PCM systems were found to be effective in keeping low module temperature, they were not found economically feasible in cold climatic conditions. So, mass production of such systems was not suggested for cold climates. However, in hot climates PV-PCM systems can be economically viable over a considerable operating period. Stropnik and Stritih [38] conducted simulation as well as experimental investigations on PV-PCM module using paraffinic organic PCM RT28HC. A 250 W solar PV module was modified to contain PCM on the back side. Simulations were carried out on TRNSYS software and a comparison of experimental and simulation data was also conducted. Annual increment of 7.3% in electrical output and 0.8% increment in conversion efficiency were reported for climatic conditions of Ljubljana, Slovenia.

Zhao et al. [39] conducted simulation study on PV-PCM system for varied PCM layer thickness. One dimensional thermal resistance model was used for performance analysis of PV-PCM system while RT35HC was used as PCM. A maximum temperature drop of 24.9 °C was achieved in one day of operation with PCM layer thickness of 30 mm (Fig. 5). Sharma et al. [40] conducted experimental study to investigate feasibility of using PCM for thermal management of low concentration building integrated concentrated photovoltaic system (BICPV). A PCM container filled with paraffin wax RT42 was integrated to the BICPV system for an indoor experimental setup. BICPV-PCM system achieved an average temperature reduction of 3.8 °C and 7.7% enhancement in electrical efficiency when compared to naturally ventilated BICPV system. Klugmann-radziemska and Wcisło-kucharek [41] conducted several experiments in laboratory conditions on PV-PCM systems in order to analyze the effectiveness of various PCMs for temperature control. Module temperature reduction of 7 °C was achieved without using active water cooling. Paraffin RT 42–44 were recommended as PCMs for PV module thermal

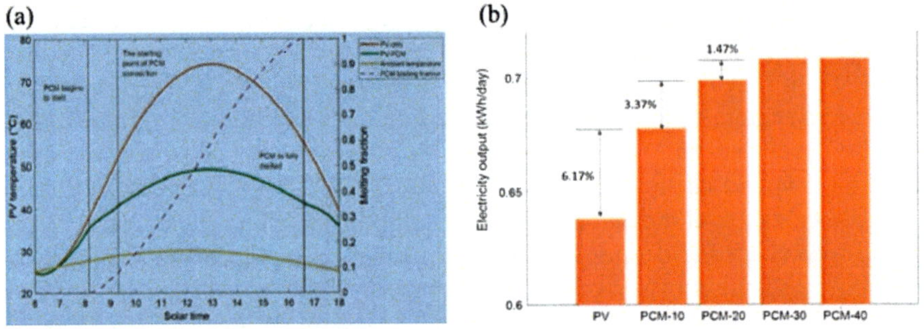

Fig. 5 **a** PV Module temperature with and without PCM system (30 mm PCM) **b** Electrical output of PV-PCM module with increasing PCM layer thickness (10, 20, 30, 40 mm) [39]

management as they were able to maintain low module temperature for most of high-intensity solar radiation period.

3 Studies on Heat Transfer Enhancement in PV-PCM Modules

Various techniques have been used by researchers in past few years to enhance the heat transfer rate of PCM to improve the performance of PV-PCM modules. Rajvikram et al. [42] conducted experimental investigation on a 5 W PV-PCM module, placing an aluminum plate at the module back as thermal conductivity enhancer. Paraffinic organic PCM OM-29 with a thickness of 30 mm was placed at the module back. The aluminum plate was found to enhance the thermal dissipation capacity of PCM. An average temperature reduction of 10.35 °C was observed during experiments with 2% improvement in overall efficiency of PV module. In order to develop an effective passive cooling system for PV modules, Bayrak et al. [19] conducted various experiments on PV-PCM modules using two different PCMs with several fin configurations (Fig. 6). Biphenyl and calcium chloride hexahydrate ($CaCl_2.6H_2O$) were used with thermoelectric modules and a number of aluminum fins on the module back. Calcium chloride hexahydrate was found to

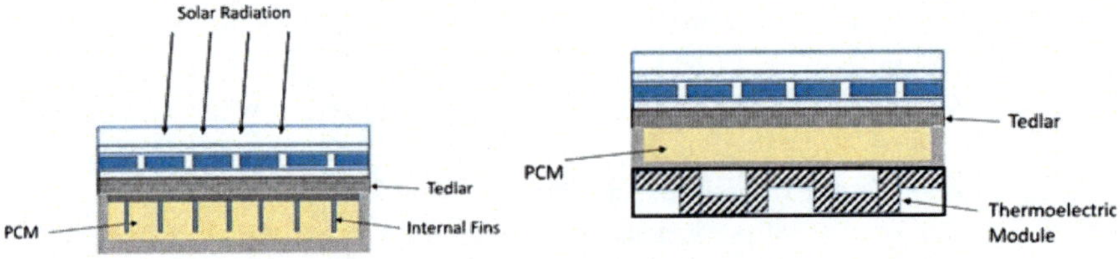

Fig. 6 Different configurations of PCM with fins and TEM

improve the PV module performance. Moreover, increasing the number of fins resulted in high-cooling performance. Highest improvement of 7.72% in PV module output was observed when $CaCl_2.6H_2O$ was used with fins and thermoelectric module .

In order to enhance thermal conductivity of paraffin wax PCM, Shastry and Arunachala [43] utilized aluminum matrix in PV/T-PCM system. Aluminum matrix, with honeycomb structure filled with paraffin wax PCM, was placed on the back side of PV module to store the heat generated by the PV module. Figure 7 shows the matrix placed in PCM container to enhance its thermal conductivity. Maximum improvement of 8.6% in module output was reported as compared to conventional PV-PCM system. Wongwuttanasatian et al. [44] utilized low-cost palm wax as PCM in three different designs of PCM container placed on PV module back. Palm wax has a melting temperature ranging from 50 to 55 °C which makes it suitable for PV module cooling. Three designs were basically three different fin configurations with provision of active water cooling. The module with finned PCM container was able to reduce the module surface temperature by 6 °C which resulted in 5.3% improvement in conversion efficiency when compared to a non-cooled PV module. Atkin and Farid [45] conducted experimental and simulation studies to investigate viability of PCM infused graphite along with external-finned heat sink for thermal management of PV modules. PV-PCM configuration which used 92% (by weight) PCM infused in graphite block and external Aluminum fins was reported to be most effective with 12.97% increase in overall efficiency of PV module.

In order to enhance thermal characteristics of PCM used in PV cooling systems, Numan [46] utilized aluminum matrix foam to modify a 67.5 W PV module into the PV-PCM module. Paraffin wax was embedded in pores of foam to store excess heat. Optimum thickness of PCM layer was found to be 3 cm at which maximum reduction in peak module temperature was achieved. Khanna et al. [47] constructed a PCM container of paraffin RT 25 HC with internal finned structure. They conducted simulation study on finned PV-PCM system to optimize several fin parameters such as fin thickness, fin length, fin spacing, and depth of PCM layer. Maiti et al. [48] used paraffin wax PCM with PV panel to limit the temperature while taking benefit of high solar intensity. Paraffin wax with melting temperature of 58 °C was incorporated at the back side of the PV collector. An attempt was made to enhance thermal conductivity of paraffin wax by spreading metal turnings

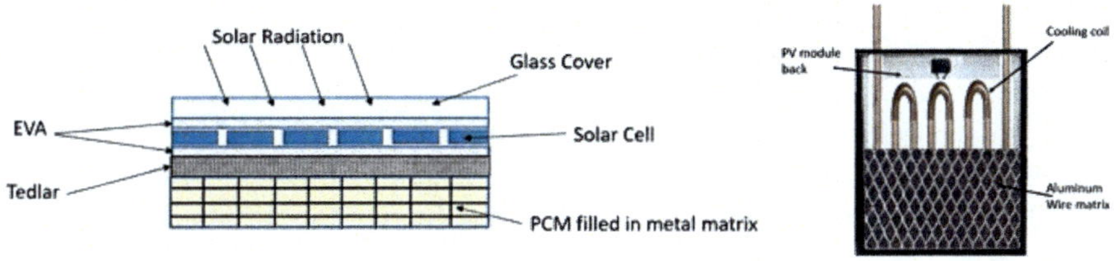

Fig. 7 PCM embedded in Al metal matrix [43]

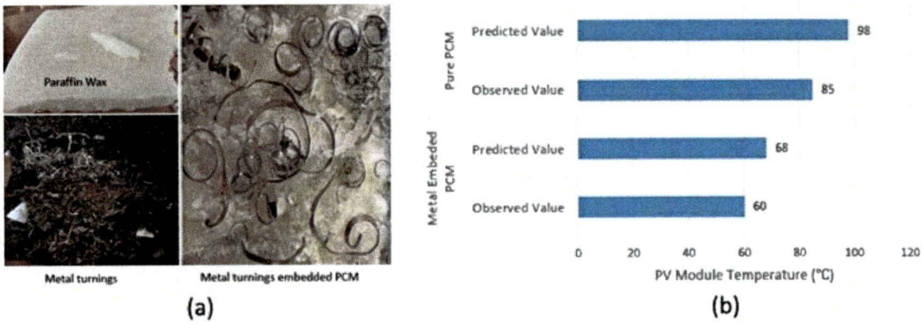

Fig. 8 **a** PCM embedded with metal turnings **b** Maximum temperature of PV module [48]

(shown in Fig. 8) in the PCM container. PCM was able to keep relatively low module temperature during indoor as well as outdoor testing of the module.

In order to test the effectiveness of activated alumina (Zeolite) as a heat storage medium, Ragab et al. [49] conducted experimental study on different configurations of PV modules. Tests were conducted in indoor conditions with four different zeolite layer thickness on back of the PV module. Metal fin structures, metal mesh and metal particles were used inside the zeolite container to further enhance the effectiveness of the system. Significant drop of 15 °C and 9 °C in module surface temperature with 10% and 7% improvement in efficiency was reported at radiation intensities of 600 W/m^2 and 1000 W/m^2, respectively. Singh et al. [50] designed a number of small aluminum containers for PCM in order to augment its thermal conductivity.

4 Studies on Composite PCMs/Nano-Enhanced PCMs

Composite PCMs which are mixture of two or more PCMs sometimes provide better thermal characteristics than a single PCM. Zhang and Zhang [51] conducted indoor experiments on composite PCM of $CaCl_2.6H_2O$ and $MgCl_2.6H_2O$ for thermal management of PV modules. Composite PCM was made by physical mixing of the two PCMs with expanded graphite as the carrier. The composite PCM showed excellent thermal characteristics and performed better than a single. The composite PCM was able to maintain PV module temperature less than 40 °C for most of the operational time which resulted in nearly 8% more electrical power output. Luo et al. [52] used a form-stable paraffin (RT28)/expanded graphite composite, to develop a PV-PCM system and conducted experiments as well as CFD simulations in order to control the temperature of PV module. Composite PCM was able to keep the temperature of module below 50 °C throughout the experiment. Figure 9 presents the findings of this study.

Siahkamari et al. [53] conducted experimental study on sheep fat utilized as PCM in water cooled PV/T system. Copper microchannels were placed on the back

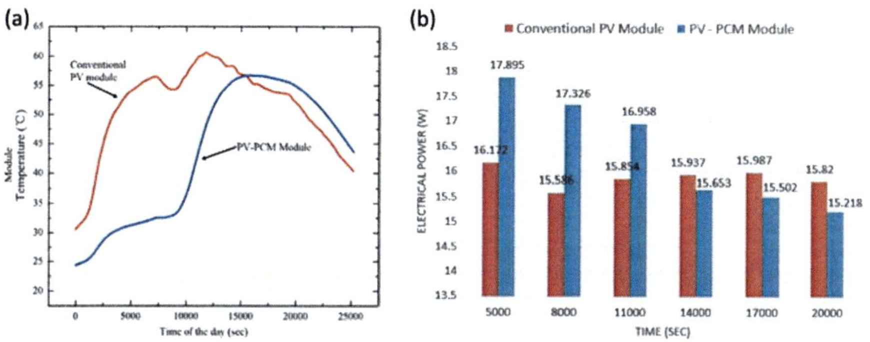

Fig. 9 **a** Temperature of conventional PV module and PV-PCM system **b** Power output comparison [52]

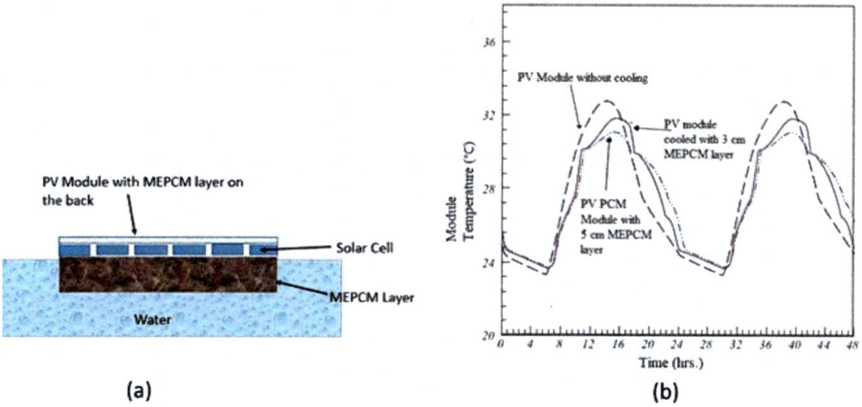

Fig. 10 **a** MEPCM layer used with floating PV module **b** Average PV module temperature with increasing MEPCM layer thickness (3 cm and 5 cm) [54]

of module through which cooling water flows. In order to improve the effectiveness of sheep fat as PCM, CuO nanoparticles were added. The nano-enhanced PV-PCM module reported 26% increment in power output when compared to PV module without cooling. Ho et al. [54] conducted numerical simulations to investigate the effect of micro-encapsulated PCM (MEPCM) layer on the surface temperature of PV module. A PV-PCM configuration was designed in which MEPCM layer was attached to the back of the module and whole system floats in water (Fig. 10). So a combined cooling effect of MEPCM and water cooling was analyzed through numerical method. Maximum reduction of 18 °C was achieved in module temperature with 5 mm thick MEPCM layer.

Tanuwijava et al. [55] conducted CFD simulation study on the use of MEPCM for temperature reduction of PV modules. A valid CFD model was developed to simulate PV module integrated with MEPCM layer on the back of the module.

Sharma et al. [30] used RT 42 as phase change material and enhanced its properties by adding nanoparticles and employing micro fins.

5 PCMs Used with Active Cooling

In order to enhance the rate of heat transfer from the module back, several researchers have used phase change materials with active cooling. Figure 11 represents a typical configuration of PV-PCM module with active cooling. Water is commonly used as coolant in case of active cooling but some authors have even used nanofluids as cooling medium. Preet et al. [56] conducted experimental studies on three different PV/T systems. First one was conventional water cooled PV/T system, second one with double absorber, and third was water cooled PV/T system with paraffin PCM. Thermal and electrical performance of all the three configurations were observed and compared with varying flow rate of water.

The PCM-based cooling system was found to be very effective in bringing down the module operating temperature and enhancing the efficiency of PV/T system. A reduction of 53% in peak module temperature was achieved with PV/T-PCM cooling system. Abood and Shahad [57] constructed aluminum pockets on the rear side of PV module with provision of coolant circulation. SiC/water nanofluid was used to remove excess heat from the module back. System was tested under outdoor conditions with different flow rates of nanofluid coolant in order to optimize the system parameters. Performance of nanofluid-cooled PV module was compared with that of conventional module under same operating conditions. Nanofluid coolant with 0.5% SiC nanoparticles at a flow rate of 2 LPM provided maximum improvement of nearly 33% in PV efficiency. Aberoumand et al. [58] developed a PV/T system in which nanofluids were circulated through several channels on module back in order to remove the excess heat. Significant improvements in thermal characteristics were observed when Ag nanoparticles were mixed in water.

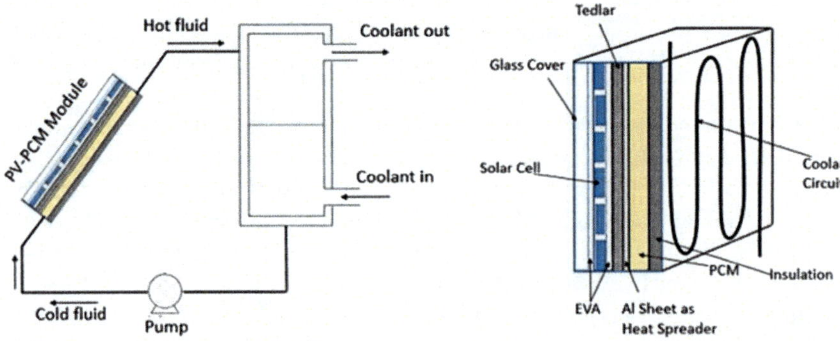

Fig. 11 Active cooling of PV modules

Increasing the concentration of Ag nanoparticles in water was found to have further improved the effectiveness of cooling system. An improvement of nearly 14% in electrical efficiency of PV module was reported with nanofluid circulation having 4% concentration of Ag nanoparticles. Nasef et al. [59] developed and modeled an integrative system for concentrated PV module thermal regulation which utilizes PCM and active water cooling. PCM RT35HC was used with different configurations of cooling fluid circulation on rear side of the module. In order to improve the overall efficiency of cooling system, CuO nanofluid with 4% concentration in water was also circulated as heat transfer fluid. CPV efficiency was increased by 2.7% with a 4 °C reduction in maximum PV temperature. Abdulmunem et al. [60] developed a passive cooling method for PV modules by employing PCM within the copper foam matrix. Phase change material was also enhanced by mixing 0.2% multi–walled carbon nanotubes. Three different configurations of PV-PCM module were developed and compared experimentally with conventional PV module. First one consisted of only a PCM on PV module back. In second design, PCM was embedded in copper foam matrix which augmented its thermal characteristics. Finally, third configuration consisted of nano-enhanced PCM embedded in copper foam matrix which was placed on module back. Improvement in conversion efficiency of PV module over that of conventional PV module in three cases were found to be 4.5%, 1.97%, and 5.68%, respectively. It was reported that copper foam matrix and carbon nanotubes are effective in improving the performance of PV modules, but their economic feasibility still remains to be explored. In order to take advantage of active and passive cooling methods at the same time, Rajaee et al. [61] integrated six different cooling methods and constructed one novel cooling system for PV modules. The novel system utilized a thermoelectric generator on module back with provision of nanofluid cooling. Paraffin wax enhanced with alumina powder was also incorporated in the integrated cooling system. The combined unit was found to be very effective in utilizing the incident solar radiation. The cooling system which utilizes nano-enhanced PCM with 1% Co_3O_4/water nanofluid as coolant improved the overall efficiency by 12.28% when compared to PV/T water system. Use of TEG further improves the effectiveness of integrated cooling system for PV modules.

6 Conclusion

Following conclusions can be drawn as per the literature survey on PCM-based cooling systems for PV modules:

- There is no single universally accepted cooling method for PV modules which can be commercialized for large PV installations. Active cooling methods are more effective than passive cooling methods, but their economic feasibility needs to be explored before employing these in practical situation.

- Phase change materials are suitable and effective in improving the efficiency of PV panels as they do not require external power input and sophisticated equipment for their operation. However, their economic feasibility could only be justified for hot climatic conditions.
- Several cooling systems employ PCMs with an active cooling system. These methods utilize benefits of active cooling as well as passive methods.
- Use of phase change materials provide many advantages but their low thermal conductivity remains a big issue. Several methods such as fins, metal matrix, and nanomaterials can be used to improve their thermal conductivity.
- Use of nanoparticles in PCM improves its thermal performance significantly, but the economic feasibility was not achieved in most of the cases.
- Mass production of PV-PCM modules is justified only for hot climatic conditions as the payback period of these systems was found to be relatively low when employed in tropical regions.

7 Future Scope

- Long-term experiments need to be conducted to evaluate the economic feasibility and durability of PV-PCM systems with seasonal variations.
- Low-cost innovative methods can be developed to address the general problem of low thermal conductivity in almost all the types of phase change materials.
- There is a need for a simple, low cost, and durable cooling system for PV panels to maintain its operating temperature within the desirable limit especially during the peak hours of solar radiation. An integration of passive and active cooling methods can be useful to achieve this.
- Various simulation studies can be conducted to establish set of standards for selection of PCM as per ambient conditions and PV cell material.
- Standard methods for performance evaluation of PV-PCM modules can be developed so that the effectiveness of a cooling system can be easily analyzed.

References

1. Elavarasan, R.M., et al.: A comprehensive review on renewable energy development, challenges, and policies of leading Indian states with an international perspective. IEEE Access **8**(May), 74432–74457 (2020). https://doi.org/10.1109/ACCESS.2020.2988011
2. Irena, Renewable Energy Statistics 2020 Statistiques D'Énergie Renouvelable 2020 Estadísticas De Energía Renovable 2020 about Irena (2020)
3. BP, Statistical Review of World Energy, 2020, 69th Edition, Bp, p. 66 (2020)
4. Solar Energy Industries Association, Solar Means Business 2015, no. October (2015)

5. Hayat, M.B., Ali, D., Monyake, K.C., Alagha, L., Ahmed, N.: Solar energy—a look into power generation, challenges, and a solar-powered future. Int. J. Energy Res. **43**(3), 1049–1067 (2019). https://doi.org/10.1002/er.4252
6. Aderhold, E.D.: Renewable energy. Archit. Dig. **68**(3), 40 (2011). https://doi.org/10.1109/ICCEP.2011.6036306
7. New and renewable energy development corporation of AP Ltd (NREDCAP), Tadepalli status of renewable energy power projects commissioned in Andhra Pradesh state as on 28. 02. 2019 Resource Cumulative capacity commissioned up to Capacity commissioned dur (2019)
8. Spelling, J., Gallo, A., Romero, M., González-Aguilar, J.: A high-efficiency solar thermal power plant using a dense particle suspension as the heat transfer fluid. Energy Procedia **69**, 1160–1170 (2015). https://doi.org/10.1016/j.egypro.2015.03.191
9. Green, M.A., Dunlop, E.D., Hohl-Ebinger, J., Yoshita, M., Kopidakis, N., Hao, X.: Solar cell efficiency tables (version 56). Prog. Photovoltaics Res. Appl. **28**(7), 629–638 (2020). https://doi.org/10.1002/pip.3303
10. Ghosh, S., Yadav, V.K., Mukherjee, V.: Impact of environmental factors on photovoltaic performance and their mitigation strategies–a holistic review. Renew. Energy Focus **28**(00), 153–172 (2019). https://doi.org/10.1016/j.ref.2018.12.005
11. Nasrin, R., Hasanuzzaman, M., Rahim, N.A.: Effect of high irradiation on photovoltaic power and energy. Int. J. Energy Res. **42**(3), 1115–1131 (2018). https://doi.org/10.1002/er.3907
12. Lazzarin, R., Noro, M.: Photovoltaic/Thermal (PV/T)/ground dual source heat pump: optimum energy and economic sizing based on performance analysis. Energy Build.**211**, 109800 (2020). https://doi.org/10.1016/j.enbuild.2020.109800
13. Jiang, J., Wang, J., Kuo, K., Su, Y., Shieh, J., Chou, J.: Analysis of the Junction Temperature and Thermal Characteristics of Photovoltaic Modules Under Various Operation Conditions, vol. 44, pp. 292–301 (2012). https://doi.org/10.1016/j.energy.2012.06.029
14. Katsumata, N., Nakada, Y., Minemoto, T., Takakura, H.: Estimation of irradiance and outdoor performance of photovoltaic modules by meteorological data. Sol Energy Mater Sol Cells **95**(1), 199–202 (2011). https://doi.org/10.1016/j.solmat.2010.01.019
15. Dubey, S., Sarvaiya, J.N., Seshadri, B.: Temperature dependent photovoltaic (PV) efficiency and its effect on PV production in the world—a review. Energy Procedia **33**, 311–321 (2013). https://doi.org/10.1016/j.egypro.2013.05.072
16. Touati, F.A., Al-Hitmi, M.A., Bouchech, H.J.: Study of the effects of dust, relative humidity, and temperature on solar PV performance in Doha: comparison between monocrystalline and amorphous PVS. Int. J. Green Energy **10**(7), 680–689 (2013). https://doi.org/10.1080/15435075.2012.692134
17. Rahman, M.M., Hasanuzzaman, M., Rahim, N.A.: Effects of various parameters on PV-module power and efficiency. Energy Convers. Manag. **103**, 348–358 (2015). https://doi.org/10.1016/j.enconman.2015.06.067
18. Gedik, E.: Experimental investigation of module temperature effect on photovoltaic panels efficiency. J. Polytech. **19**(4), 569–576 (2016)
19. Bayrak, F., Oztop, H.F., Selimefendigil, F.: Experimental study for the application of different cooling techniques in photovoltaic (PV) panels. Energy Convers. Manag. **212**(March), 112789 (2020). https://doi.org/10.1016/j.enconman.2020.112789
20. Al-Waeli, A.H.A., Chaichan, M.T., Sopian, K., Kazem, H.A., Mahood, H.B., Khadom, A.A.: Modeling and experimental validation of a PVT system using nanofluid coolant and nano-PCM. Sol. Energy **177**(July 2018), 178–191 (2019). https://doi.org/10.1016/j.solener.2018.11.016
21. Indartono, Y.S., Suwono, A., Pratama, F.Y.: Improving photovoltaics performance by using yellow petroleum jelly as phase change material. Int. J. Low-Carbon Technol. **11**(3), 333–337 (2016). https://doi.org/10.1093/ijlct/ctu033
22. Choubineh, N., Jannesari, H., Kasaeian, A.: Experimental study of the effect of using phase change materials on the performance of an air-cooled photovoltaic system. Renew. Sustain. Energy Rev. **101**(October 2018), 103–111 (2019). https://doi.org/10.1016/j.rser.2018.11.001

23. Waqas, A., Ji, J., Bahadar, A., Xu, L., Modjinou, M.: Thermal Management of Conventional Photovoltaic Module Using Phase Change Materials—An Experimental Investigation (2019). https://doi.org/10.1177/0144598718795697
24. Hasan, A., Alnoman, H., Shah, A.H.: Energy efficiency enhancement of photovoltaics by phase change materials through thermal energy recovery. Energies **9**(10) (2016). https://doi.org/10.3390/en9100782
25. Hossain, M.S., Pandey, A.K., Selvaraj, J., Rahim, N.A., Islam, M.M., Tyagi, V.V.: Two side serpentine flow based photovoltaic-thermal-phase change materials (PVT-PCM) system: energy, exergy and economic analysis. Renew. Energy **136**, 1320–1336 (2019). https://doi.org/10.1016/j.renene.2018.10.097
26. Sarafraz, M.M., et al.: Experimental investigation on thermal performance of a PV/T-PCM (photovoltaic/thermal) system cooling with a PCM and nanofluid. Energies **12**(13), 1–16 (2019). https://doi.org/10.3390/en12132572
27. Abdollahi, N., Rahimi, M.: Potential of water natural circulation coupled with nano-enhanced PCM for PV module cooling. Renew. Energy **147**, 302–309 (2020). https://doi.org/10.1016/j.renene.2019.09.002
28. Al-Waeli, A.H.A., et al.: Comparison study of indoor/outdoor experiments of a photovoltaic thermal PV/T system containing SiC nanofluid as a coolant. Energy **151**, 33–44 (2018). https://doi.org/10.1016/j.energy.2018.03.040
29. Eisapour, M., Hossein, A., Hosseini, M.J., Talebizadehsardari, P.: Exergy and energy analysis of wavy tubes photovoltaic-thermal systems using microencapsulated PCM nano-slurry coolant fluid. Appl. Energy **266**(December 2019), 114849 (2020). https://doi.org/10.1016/j.apenergy.2020.114849
30. Sharma, S., Micheli, L., Chang, W., Tahir, A.A., Reddy, K.S., Mallick, T.K.: Nano-enhanced phase change material for thermal management of BICPV. Appl. Energy, no. September 2017, 1–15 (2020). https://doi.org/10.1016/j.apenergy.2017.09.076
31. Salem, M.R., Elsayed, M.M., Abd-Elaziz, A.A., Elshazly, K.M.: Performance enhancement of the photovoltaic cells using Al2O3/PCM mixture and/or water cooling-techniques. Renew. Energy **138**, 876–890 (2019). https://doi.org/10.1016/j.renene.2019.02.032
32. Rajvikram, M., Sivasankar, G.: Experimental study conducted for the identification of best heat absorption and dissipation methodology in solar photovoltaic panel. Sol. Energy **193** (March), 283–292 (2019). https://doi.org/10.1016/j.solener.2019.09.053
33. Mahamudul, H., et al.: Temperature Regulation of Photovoltaic Module Using Phase Change Material : A Numerical Analysis and Experimental Investigation, vol. 2016 (2016)
34. Hasan, A., Sarwar, J., Alnoman, H., Abdelbaqi, S.: Yearly energy performance of a photovoltaic-phase change material (PV-PCM) system in hot climate. Sol. Energy **146**, 417–429 (2017). https://doi.org/10.1016/j.solener.2017.01.070
35. Kawtharani, F., et al.: Cooling PV modules using phase change material. Proc. Int. Conf. Microelectron. ICM **2017-Decem**(Icm), 1–5 (2018). https://doi.org/10.1109/ICM.2017.8268830
36. Hendricks, J.H.C., Van Sark, W.G.J.H.M.: Annual performance enhancement of building integrated photovoltaic modules by applying phase change materials. Prog. Photovoltaics Res. Appl. **21**(4), 620–630 (2013). https://doi.org/10.1002/pip.1240
37. Hasan, A., McCormack, S.J., Huang, M.J., Norton, B.: Energy and cost saving of a photovoltaic-phase change materials (PV-PCM) System through temperature regulation and performance enhancement of photovoltaics. Energies **7**(3), 1318–1331 (2014). https://doi.org/10.3390/en7031318
38. Stropnik, R., Stritih, U.: Increasing the efficiency of PV panel with the use of PCM. Renew. Energy **97**, 671–679 (2016). https://doi.org/10.1016/j.renene.2016.06.011
39. Zhao, J., Li, Z., Ma, T.: Performance analysis of a photovoltaic panel integrated with phase change material. Energy Procedia **158**, 1093–1098 (2019). https://doi.org/10.1016/j.egypro.2019.01.264

40. Sharma, S., Tahir, A., Reddy, K.S., Mallick, T.K.: Solar energy materials & solar cells performance enhancement of a building-integrated concentrating photovoltaic system using phase change material. Sol. Energy Mater. Sol. Cells **149**, 29–39 (2016). https://doi.org/10.1016/j.solmat.2015.12.035
41. Klugmann-radziemska, E., Wcisło-kucharek, P.: Photovoltaic module temperature stabilization with the use of phase change materials. Sol. Energy **150**, 538–545 (2017). https://doi.org/10.1016/j.solener.2017.05.016
42. Rajvikram, M., Leoponraj, S., Ramkumar, S., Akshaya, H., Dheeraj, A.: Experimental investigation on the abasement of operating temperature in solar photovoltaic panel using PCM and aluminium. Sol. Energy **188**(February), 327–338 (2019). https://doi.org/10.1016/j.solener.2019.05.067
43. Shastry, D.M.C., Arunachala, U.C.: Thermal management of photovoltaic module with metal matrix embedded PCM. J. Energy Storage **28**(February), 101312 (2020). https://doi.org/10.1016/j.est.2020.101312
44. Wongwuttanasatian, T., Sarikarin, T., Suksri, A.: Performance enhancement of a photovoltaic module by passive cooling using phase change material in a finned container heat sink. Sol. Energy **195**(July 2019), 47–53 (2020). https://doi.org/10.1016/j.solener.2019.11.053
45. Atkin, P., Farid, M.M.: Improving the efficiency of photovoltaic cells using PCM infused graphite and aluminium fins. Sol. Energy **114**, 217–228 (2015). https://doi.org/10.1016/j.solener.2015.01.037
46. Numan, A.H.: An Experimental Investigation of Performance of Hybrid PV / T-PCM System with Control System An Experimental Investigation of Performance of Hybrid PV/T-PCM System with Control System (2020). https://doi.org/10.1088/1757-899X/765/1/012049
47. Khanna, S., Reddy, K.S., Mallick, T.K.: International Journal of Thermal Sciences Optimization of fi nned solar photovoltaic phase change material (fi nned pv pcm) system, vol. 130, no. May, pp. 313–322 (2018)
48. Maiti, S., Banerjee, S., Vyas, K., Patel, P., Ghosh, P.K.: Self Regulation of Photovoltaic Module Temperature in V-trough Using a Metal—Wax Composite Phase Change Matrix, vol. 85, pp. 1805–1816 (2011). https://doi.org/10.1016/j.solener.2011.04.021
49. Ragab, S., Saidani-scott, H., Benedi, J.: Experimental study for thermal regulation of photovoltaic panels using saturated zeolite with water. Sol. Energy **188**(April), 464–474 (2019). https://doi.org/10.1016/j.solener.2019.06.039
50. Singh, P., Mudgal, V., Khanna, S., Mallick, T.K., Reddy, K.S.: Experimental investigation of solar photovoltaic panel integrated with phase change material with multiple conductivity-enhancing-containers. Energy, 118047 (2020). https://doi.org/10.1016/j.energy.2020.118047
51. Zhang, Y., Zhang, X.: Thermal properties of a new type of calcium chloride hexahydrate-magnesium chloride hexahydrate/expanded graphite composite phase change material and its application in photovoltaic heat dissipation. Sol. Energy **204**(May), 683–695 (2020). https://doi.org/10.1016/j.solener.2020.05.037
52. Luo, Z., et al.: Numerical and experimental study on temperature control of solar panels with form-stable paraffin/expanded graphite composite PCM. Energy Convers. Manag. **149**(July), 416–423 (2017). https://doi.org/10.1016/j.enconman.2017.07.046
53. Siahkamari, L., Rahimi, M., Azimi, N., Banibayat, M.: Experimental investigation on using a novel phase change material (PCM) in micro structure photovoltaic cooling system. Int. Commun. Heat Mass Transf. **100**(January), 60–66 (2019). https://doi.org/10.1016/j.icheatmasstransfer.2018.12.020
54. Ho, C.J., Chou, W.L., Lai, C.M.: Thermal and electrical performance of a water-surface floating PV integrated with a water-saturated MEPCM layer. Energy Convers. Manag. **89**, 862–872 (2015). https://doi.org/10.1016/j.enconman.2014.10.039
55. Tanuwijava, A.O., Ho, C.J., Lai, C.M., Huang, C.Y.: Numerical investigation of the thermal management performance of MEPCM modules for PV applications. Energies **6**(8), 3922–3936 (2013). https://doi.org/10.3390/en6083922

56. Preet, S., Bhushan, B., Mahajan, T.: Experimental investigation of water based photovoltaic/ thermal (PV/T) system with and without phase change material (PCM). Sol. Energy **155**, 1104–1120 (2017). https://doi.org/10.1016/j.solener.2017.07.040
57. Abood, M.H., Shahad, H.: Improving the Electrical Efficiency of a Photovoltaic/Thermal Panel by Using SiC/Water Nanofluid as Coolant Improving the Electrical Efficiency of a Photovoltaic/Thermal Panel by Using SiC/Water Nanofluid as Coolant, no. January (2020). https://doi.org/10.1088/1757-899X/671/1/012143
58. Aberoumand, S., Ghamari, S., Shabani, B.: Energy and exergy analysis of a photovoltaic thermal (PV/T) system using nano fluids: An experimental study. Sol. Energy **165**(March), 167–177 (2018). https://doi.org/10.1016/j.solener.2018.03.028
59. Nasef, H.A., Nada, S.A., Hassan, H.: Integrative passive and active cooling system using PCM and nano fl uid for thermal regulation of concentrated photovoltaic solar cells. Energy Convers. Manag. **199**(September), 112065 (2019). https://doi.org/10.1016/j.enconman.2019.112065
60. Abdulmunem, A.R., Mohd, P., Abdul, H., Hussien, H.A., Izmi, I.: Enhancing PV Cell's electrical efficiency using phase change material with copper foam matrix and multi-walled carbon nanotubes as passive cooling method. Renew. Energy **160**, 663–675 (2020). https://doi.org/10.1016/j.renene.2020.07.037
61. Rajaee, F., Rad, M.A.V., Kasaeian, A., Mahian, O., Yan, W.M.: Experimental analysis of a photovoltaic/thermoelectric generator using cobalt oxide nanofluid and phase change material heat sink. Energy Convers. Manag. **212**(March), 112780 (2020). https://doi.org/10.1016/j.enconman.2020.112780

Thermal Comfort and Energy Efficient Design of a Central Air Conditioning System of an Educational Institute

Rahul Sharma, Roshan Raman, Omprakash Yadav, Mohammad Shahbaz Khan, and Vinod Kumar Yadav

1 Introduction

Today most of the world in country is experiencing global warming and the effect of this result in the form of increasing temperature year by year in the summer period. And as a result, the energy required fulfilling the demand of air conditioning also increases yearly [1]. It has a big and challenging task to provide technologies which is energy efficient [1] to fulfill the growing demand with minimum impact on environment. Efficiency [1] is very important in central air conditioning system as it determines the amount of energy that is being consumed for heating and cooling. The central air conditioning system considers [2] the most modern AC systems and efficient energy usage. Central AC system has some advantage over other conditioning system; some of them are consumption of less energy with better life, and another most favorable advantage is it fulfill the cooling requirement in all the space of the building at the same time [2]. Therefore, it becomes the need of commercial building like educational institution building. Central air conditioning system requires lower power than other AC system serves multiple spaces from one base location. The central air conditioning system [3] is working on the same principle as in the normal unit of air conditioning system. It works on the basic refrigeration cycle, i.e., vapor compression refrigeration cycle.

R. Sharma · M. S. Khan (✉) · V. K. Yadav
Department of Mechanical Engineering, G.L. Bajaj Institute of Technology and Management, Greater Noida, Uttar Pradesh 201308, India

R. Raman · O. Yadav
Department of Mechanical Engineering, The NorthCap University, Gurugram 122017, India

L. M. Das et al. (eds.), *Recent Trends in Thermal Engineering*, Lecture Notes in Mechanical Engineering,
https://doi.org/10.1007/978-981-16-3428-4_12

2 Research Methodology

The following research design methodology has been utilized for the central air conditioning design of the building:

A. **Effective Zoning System**
   The idea of a zoning system [1] is very effective which allow each zone or room has its independent control. Each zone has its own thermostat for controlling temperature and motorized damper in the duct to regulate and redirect cooled air to specific areas of the space. Zoning system has multiple benefits like save money, comfort, longer system life, etc. Here, the complete space of the building was specified in two zones, and each zone was operated with multiple air handling unit (AHU) and fan coil unit (FCU) as shown in Fig. 2.
B. **Waste-Heat Recovery**
   Waste-heat recovery is very important for the conservation of energy. Heat recovery devices used to recover thermal energy from exhaust air and transfer it to the incoming fresh-air supply. This gives the reduction in the energy that is normally required to heat or cool air of the system. A good heat recovery device saves approx. 10% of the running cost of the HVAC system.

3 Experimental Investigation

This section determines the total heat load estimation for determining actual tonnage or amount of heat generated in the building space. This section also describes about proper equipment selection like chiller, FCU, AHU, and some air distribution and control product like diffuser and damper. It also includes major part of the central AC system, i.e., chilled water pipe sizing, duct sizing, and drafting.

3.1 Heat Load Estimation

Heat load estimation includes building survey, psychrometric analysis, and load calculation.

3.1.1 Building Survey of the Building

There are some necessary details [6] to design the central air conditioning system like building orientation, building spacing, physical dimensions of the space, ceiling height, floor-to-floor height, space above the false ceiling, etc. It includes all

Table 1 Dimensions of the building

1. Total floor	10
2. Total area of the building	104,837 sq ft
3. Ceiling height	10 ft
4. Space above the false ceiling	2 ft
5. Exterior wall of the building	499 mm
6. Interior wall of the building	256 mm
7. Thickness of roof	165 mm

the details that required proceeding for load calculation. The dimension of the building is given below as shown in Table 1.

Orientations of the building are north-east, north-west, south-east, and south-west.

3.1.2 Design Condition

This analysis gives the idea of as which psychometric process needs to be done as outside condition and inside requirement of the building as shown in Table 2.

3.1.3 Load Calculation

The objective of cooling load calculation is to find the right selection of air conditioning equipment to achieve comfort operation and good air distribution in the air-conditioned zone. Major components of loads in the building are:

Heat Transfer Through Building Structure

There [2] are various way of heat transfer in the building which to be considered in the air conditioning of building, i.e., heat transfer through walls, roofs, ceiling, floor, window, etc.

There are two methods to calculate the heat transfer through walls and the roof, but equivalent temperature differential method is commonly used by HVAC design

Table 2 Design conditions of the building

	Inside condition	Outside condition
DBT (°F)	75	110
WBT (°F)	62.5	75
RH%	50	20
DPT (°F)	56	60
HR	65	78

engineers, as it is applicable to sunlight walls and roofs. According to this method, the heat transfer is given by:

$$Q = U \times A \times (T_2 - T_1)$$

where Q = heat flow rate$\left(\frac{\text{KJ}}{\text{Sec}}\right)$

U = transmission rate.

A = Area of surface (Square meter).

$(T_2 - T_1)$ = Equivalent temperature difference.

With the help of ASHRAE, data note 2005 [10] can calculate equivalent temperature difference.

Transmission Heat Gain Through Glass

This is heat gain that is obtained due to the temperature difference (Dry bulb temperature) in outside and inside conditions of the building structure. Here area of all glasses exposed to the sun taken into consideration.

Occupancy Load

Heat is produced within the human body by oxidation called metabolic rate. The amount of heat dissipated by the human body by radiation and convection is determined by the temperature difference between the body surface and its surrounding. The metabolic rate is 85% for the male, and for children, it is about 75%.

Lighting

Light [2] produces heat, and the heat is dissipated by radiation to the surrounding surfaces, by conduction into the adjacent materials and by convection to the surrounding air.

Formulae to be used to calculate the lighting load is

$$\text{Total lighting load} = \text{LPD} \times \text{floor area} \times \text{FOS}$$

Whereas

LPD = lighting power density (1.24).

FOS = Factor of safety = 1.25.

Appliances

Most applications are responsible for sensible as well as latent heat to a space. Electric appliances contribute latent heat only by drying, cooking, etc., whereas gas burning appliances contribute additional moisture as a product of combustion.

Heat load calculation method

Generally, heat load can be calculated by two methods one is manual calculation through E-20 sheet and other one is software calculation, i.e., calculation by hourly analysis program (HAP) software. Here E-20 sheet for the heat load calculation. Sample of E-20 sheet is given in Fig. 1.

Heat load of the Building

From the entire floor of the building, the total heat load obtained 650 TR and 234,743 dehumidified CFM.

3.2 Equipment Selection

The selection of machine, i.e., chiller, AHU, FCU, diffuser, grill, and dampers is very important for an efficient central air conditioning system.

3.2.1 Chillers

All buildings [4] need air conditioning, and there is need to remove heat to maintain the comfort temperature inside the building. Centralized systems use chillers which produce chilled water which is used to provide air conditioning in buildings. The idea behind using chiller in large building is cost effective.

Performance rating of a chiller

I. COP (coefficient of performance) = cooling capacity required (kW)/Total power input (kW)

Total power input

$$\text{Air cooled chiller} = \text{compressor power input} + \text{condenser fan}$$

$$\text{Water cooled chiller} = \text{compressor power input}$$

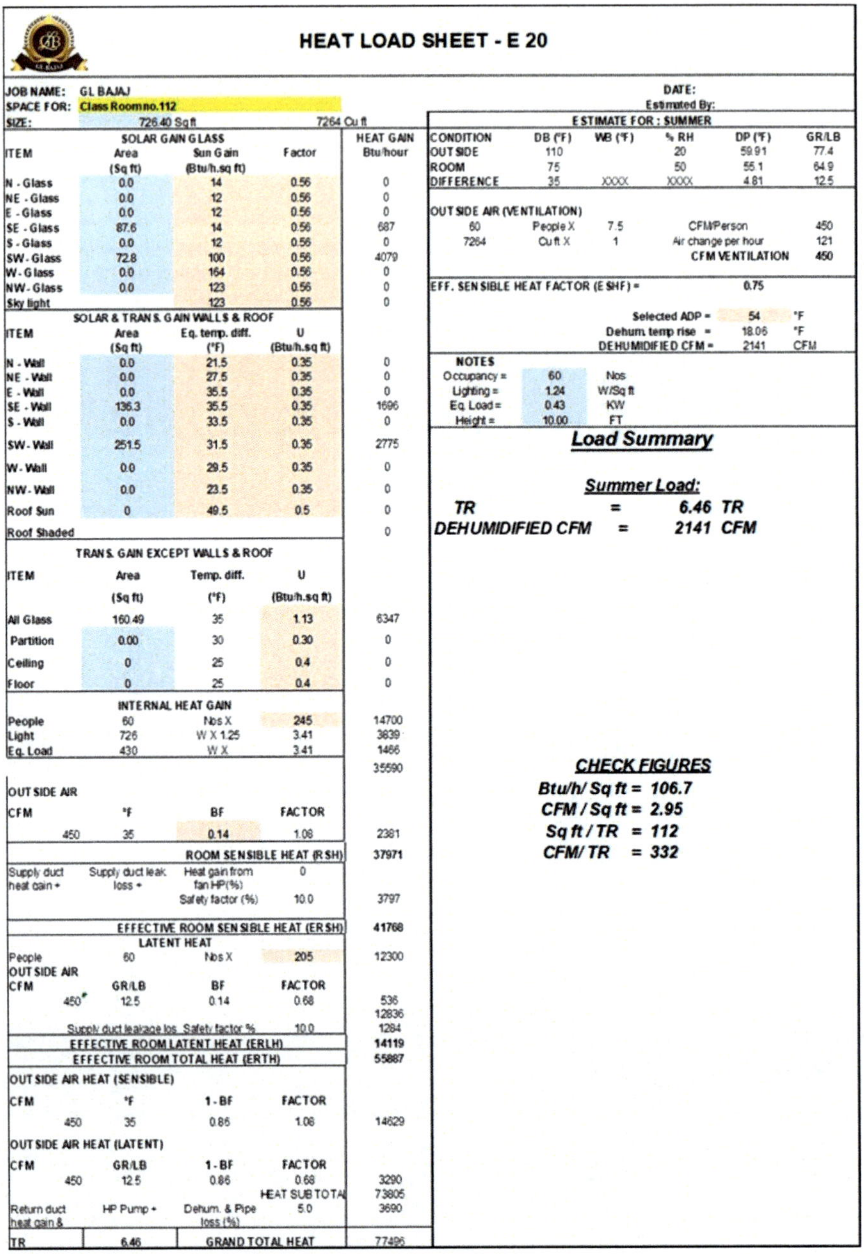

HEAT LOAD SHEET - E 20

JOB NAME: GL BAJAJ
SPACE FOR: Class Roomno.112
SIZE: 726.40 Sqft 7264 Cu ft

DATE:
Estimated By:

ITEM	Area (Sq ft)	Sun Gain (Btu/h.sq ft)	Factor	HEAT GAIN Btu/hour
SOLAR GAIN GLASS				
N - Glass	0.0	14	0.56	0
NE - Glass	0.0	12	0.56	0
E - Glass	0.0	12	0.56	0
SE - Glass	87.6	14	0.56	687
S - Glass	0.0	12	0.56	0
SW - Glass	72.8	100	0.56	4079
W - Glass	0.0	164	0.56	0
NW - Glass	0.0	123	0.56	0
Sky light		123	0.56	0

ITEM	Area (Sq ft)	Eq. temp. diff. (°F)	U (Btu/h.sq ft)	HEAT GAIN Btu/hour
SOLAR & TRANS. GAIN WALLS & ROOF				
N - Wall	0.0	21.5	0.35	0
NE - Wall	0.0	27.5	0.35	0
E - Wall	0.0	35.5	0.35	0
SE - Wall	136.3	35.5	0.35	1696
S - Wall	0.0	33.5	0.35	0
SW - Wall	251.5	31.5	0.35	2775
W - Wall	0.0	29.5	0.35	0
NW - Wall	0.0	23.5	0.35	0
Roof Sun	0	49.5	0.5	0
Roof Shaded				0

ITEM	Area (Sq ft)	Temp. diff. (°F)	U (Btu/h.sq ft)	
TRANS. GAIN EXCEPT WALLS & ROOF				
All Glass	160.49	35	1.13	6347
Partition	0.00	30	0.30	0
Ceiling	0	25	0.4	0
Floor	0	25	0.4	0

INTERNAL HEAT GAIN				
People	60	Nos X	245	14700
Light	726	W X 1.25	3.41	3639
Eq. Load	430	W X	3.41	1466
				35590

OUT SIDE AIR				
CFM	°F	BF	FACTOR	
450	35	0.14	1.08	2381
ROOM SENSIBLE HEAT (RSH)				37971
Supply duct heat gain +	Supply duct leak loss +	Heat gain from fan HP(%)	0	
		Safety factor (%)	10.0	3797
EFFECTIVE ROOM SENSIBLE HEAT (ERSH)				41768

LATENT HEAT				
People	60	Nos X	205	12300
OUT SIDE AIR				
CFM	GR/LB	BF	FACTOR	
450	12.5	0.14	0.68	536
				12836
Supply duct leakage los	Safety factor %		10.0	1284
EFFECTIVE ROOM LATENT HEAT (ERLH)				14119
EFFECTIVE ROOM TOTAL HEAT (ERTH)				55887

OUT SIDE AIR HEAT (SENSIBLE)				
CFM	°F	1 - BF	FACTOR	
450	35	0.86	1.08	14629
OUT SIDE AIR HEAT (LATENT)				
CFM	GR/LB	1 - BF	FACTOR	
450	12.5	0.86	0.68	3290
			HEAT SUB TOTAL	73806
Return duct heat gain &	HP Pump +	Dehum. & Pipe loss (%)	5.0	3690
TR	6.46	GRAND TOTAL HEAT		77496

ESTIMATE FOR : SUMMER

CONDITION	DB (°F)	WB (°F)	% RH	DP (°F)	GR/LB
OUT SIDE	110		20	59.91	77.4
ROOM	75		50	55.1	64.9
DIFFERENCE	35	XXXX	XXXX	4.81	12.5

OUT SIDE AIR (VENTILATION)				
60	People X	7.5	CFM/Person	450
7264	Cu ft X	1	Air change per hour	121
			CFM VENTILATION	450

EFF. SENSIBLE HEAT FACTOR (ESHF) = 0.75

Selected ADP =	54	°F
Dehum. temp rise =	18.06	°F
DEHUMIDIFIED CFM =	2141	CFM

NOTES

Occupancy =	60	Nos
Lighting =	1.24	W/Sq ft
Eq. Load =	0.43	KW
Height =	10.00	FT

Load Summary

Summer Load:

TR = 6.46 TR
DEHUMIDIFIED CFM = 2141 CFM

CHECK FIGURES

Btu/h/ Sq ft = 106.7
CFM / Sq ft = 2.95
Sq ft / TR = 112
CFM/ TR = 332

Fig. 1 Sample of E-20 sheet

II. Input KW/TR (IKW/TR)

$$\text{IKW/TR} = \text{Total power input (kW)/cooling capacity (TR)}$$

III. Integrated Part Load Value or Non-Standard Part Load Value

$$\text{IPLV or NPLV} = 0.01A + 0.42B + 0.45C + 0.12D$$

A = COPR OR EER AT 100% load.
B = COPR OR EER AT 75% load.
C = COPR OR EER AT 50% load.
D = COPR OR EER AT 25%. load.
IPLV used for standard condition, whereas NPLV used for site condition.

Best chiller machine in terms of performance—COP will be higher, IKW/TR will be lower, and IPLV, NPLV will be higher. Chiller selection is done on the basis of heat load calculation, performance rating of chiller and availability of resource. Capacity of 650 TR water cooled centrifugal chiller is selected. The refrigerant used in the chiller is R-1233zd (E). The total cost installation of chiller system is $300 per ton.

3.2.2 AHU (Air Handling Unit)

AHU selection is very important in the execution stage. Parameters which are the selection criteria with some recommended factor of safety as per application.

1. Heat load calculation
2. Air flow CFM
3. Total external static pressure
4. Kind of zoning system (single/multi zone)

Proper selection of fan, filter, cooling coil, and dehumidifier increases the performance rating, efficiency which results in the lower cost cycle for an application. There are two AHUS used at each floor of the building. AHU used for the class rooms, laboratories, workshop rooms, computer labs, design center rooms, seminar halls, corridor, and for entrance lobbies. Selection of AHU of following specifications, i.e., a ceiling suspended having capacity range of 200 CFM to 50,000 CFM, and the body material is aluminum.

3.2.3 FCU (Fan Coil Unit)

FCU is selected only for faculty room, HOD office, and clerk office. A ceiling suspended FCU, having capacity range of 1–3.5 ton and 400–1400 CFM, is selected for the smooth operation of air conditioning. The static pressure of FCU is 10–15 mmwc and material is GI (white color).

3.2.4 Air Distribution Product

Air distribution product is used to maintain proper combination of temperature, humidity, and air motion in the occupied zone which is normally at 1.8 m above the floor level. Diffuser [5] is installed where there is grid ceiling, and difference between one diffuser to another diffuser is 1800 to 2400 mm. If there is grid ceiling irrespective of dehumidified air quantity, there will be use of 24″ × 24″ grid size diffuser. Avoid supply diffuser near door and window.

3.2.5 Air Control Product

For proper air distribution, air control product like different types of damper is used. Dampers are used to control air because it offers low cost and low maintenance. These dampers are installed in between the duct or at inlet or outlet of collar. Some of the dampers are discussed below.

Volume Control Damper (VCD)

It is installed to control the flow of air. Size of volume control damper is equal to the size of duct. It is placed where there is network distribution.

Fire Damper

Fire dampers are used to prevent the spread of fire inside the duct. A fire damper works when the normal temperature of a room to rise to about 165 °F. It is installed in supply and return of AHU mouth. Motorized fire damper (actuator based) is used nowadays.

3.3 Design of Chilled Water Piping

Design of chilled water piping [9] is very important for central air conditioning system because it can affect the cooling of the system. Chilled water piping is that piping which joins AHU with chiller. Chiller uses the refrigeration process to chill water in a chiller barrel. This water is pumped throughout the building. Welded black steel pipe (low carbon steel) is used for the sizing of chilled water, and the medium duty grade pipe or schedule 40 is used.

3.3.1 Methods of Chilled Water Pipe Designing

Velocity and friction method is used for sizing of chilled water pipe. The condition for this method is.

1. Velocity method
   If diameter $\leq$ 2, velocity V $\leq$ 4 FPS.
2. Friction method
   If diameter > 2, friction F $\leq$ 4 ft/100 ft.

3.3.2 Parameter Required for Chilled Water Pipe Designing

Following are the parameter which required in calculating pipe sizing.

1. Flow rate in GPM (lit/sec, m3/sec)

$$1\,\text{Gallon} = 3.78\,\text{liter}$$

If (ΔT = chilled supply water temp – chilled return water temp)

$$\Delta T = 55\,^{\circ}\text{F} - 45\,^{\circ}\text{F} = 10\,^{\circ}\text{F}$$

a. 1 TR = 2.4 GPM
b. Chilled water pipe layout
c. Velocity in FPS
d. Friction (ft/100 ft or pipe length)
e. Standard chart for manual calculation
f. Pipe sizer for software calculation

3.3.3 Chilled Water Pipe Sizing of Ground Floor

Here it is impossible to present calculation of all floor of the building. So, we have presented only calculation of ground floor as a sample to understand the calculation of chilled water piping as shown in Table 3.

3.4 Duct Design of Building

Duct design [10] is very important in central air conditioning system because a good duct design increases the efficiency level at a higher stage. Ducts have improper

Table 3 Chilled water pipe data for ground floor

Section	Flow rate (GPM)	Velocity (FPS)	Head loss (ft/100 ft)	Dia (inch)
A-B	1440	9.24	3.158	8
B-1	117.48	5.1	3.299	3
1–2	2.592	1.56	1.985	0.75
1–3	114.888	4.99	3.164	3
3–8	55.2	3.7	2.376	2.5
8–9	1.512	1.6	2.983	0.5
8–10	53.688	3.6	2.257	2.5
3–4	57.6	3.86	2.57	2.5
4–7	3.6	1.34	1.105	1
4–5	54	3.62	2.281	2.5

Table 4 Recommended duct velocity for low and high velocity system (FPM)

Low velocity		High velocity	
Main duct	1000–2400	Main duct	2500–4500
Branch duct	600–1600	Branch duct	2000–4000
Return duct	1500–1800	Return duct	1800–2200

designed result in human discomfort, higher energy cost, bad air quality, and increased noise level. Duct should be considered first when new building is design.

3.4.1 Velocity of Air in Duct

Low velocity is used for commercial purpose, whereas high velocity is used for industrial purpose. Table 4 represents the range of velocity in main duct, branch duct, and return duct.

3.4.2 Pressure of Air in Duct

This pressure is total pressure including the losses through the air handling apparatus, ductwork, and the air terminal in the space.

1. Medium pressure used for primary air ductwork (fan connections, risers, and main distribution ducts).
2. Low-pressure used for secondary air ductwork (run-outs/branches from main to terminal boxes and distribution devices).

3.4.3 Duct Shape and Material

Rectangular duct is used because rectangular shape duct fit better to building structure as fit above ceilings and also easy to install between studs and joints. An aspect ratio close to 1 is the most efficient rectangular duct shape. The ideal aspect ratio for saving of duct cost of material and fan energy is 2 to 3. Galvanized steel is used for fabrication of ductwork. It saves the money and provides comfort level at higher stage. The specifications for galvanized steel sheet are ASTM A653, coating G90.

3.4.4 Methods of Duct Design

The equal friction method is used for duct design. In equal friction method, for entire system, the pressure loss per foot is same. It is not used for higher pressure system. Basically, it is used for lower pressure supply air, return air, and exhaust air duct systems. It reduces the air velocity in the direction of airflow in the supply air duct system which reduces the possibility of generating noise. This method reduced complexity of calculation and reduced time consumption.

Important formulae and point related to equal friction method

1. The friction loss will vary from 0.06″ WATER/100 FT TO 0.08″ WATER/ 100 FEET.
2. The main and branch ducts ($\Delta Pf/L$) have same frictional loss per foot.

Important consideration for the duct design

1. There should be least number of bents.
2. Air should be delivering directly as possible to economize on power, material, and space.
3. There should be permissible limits of air velocity in the duct to minimize noise.
4. Dampers should be provided in all branch outlet of the duct.
5. Remove duct obstructions.
6. Rectangular ducts should be made as square shape as possible. It ensures minimum duct surface and hence cost for the same air carrying capacity.
7. To minimize the pressure loss, there should be use of turning vanes for bends.
8. An aspect ratio of 4:1 should be conserving in the duct.

3.4.5 Duct Sizing Calculation

Tables 5 and 6 represent the duct sizing calculation of ground floor. This calculation is done by using McQuay duct sizer software.

Table 5 Duct sizing of ground floor

Section	CFM	Velocity (FPM)	Friction (In wc/wft)
P1-P2	9572	1385	0.065
P2-P3	2845	1033	0.065
P2-P4	6406	1260	0.065
P4-P6	1602	899.3	0.065
P4-P8	1602	899.3	0.065
P4-P5	3202	1065.2	0.065
P5-P7	1601	899.2	0.065
P5-P9	1601	899.2	0.065
S1-S2	8309	1339.4	0.065
S2-S9	2082	957.1	0.065
S2-S3	5473	1211	0.065
S3-S6	1557	891.2	0.065
S3-S4	3916	1116.6	0.065
S4-S8	1846	929.3	0.065
S4-S5	2070	955.7	0.065

Table 6 Required dimension of rectangular duct

Rectangular duct			
Length	Width (inch)	Height (inch)	Aspect ratio
6.394	57	20	2.85
15.095	31	14	2.07
7.120	49	17	2.88
8.015	29	10	2.9
8.015	29	10	2.9
6.536	38	13	2.92
8.015	29	10	2.9
8.015	29	10	2.9
10.245	53	19	2.78
8.865	29	12	2.41
2.467	46	16	2.87
8.263	29	10	2.9
5.839	41	14	2.92
8.263	29	11	2.63
9.725	32	11	2.90

Static pressure

Static pressure [8] is calculated for AHU selection. Internal static pressure is produced inside equipment (filter, mixing chamber, cooling coil). External static pressure is produced outside the equipment, i.e., in straight duct and in standard equipment (fittings and damper).

Internal static pressure

Total internal static pressures inside the equipment are given below

$$\text{Pre fine filter} = 5 - 7 \text{ mm of WC (water column)}$$

$$\text{Fine filter} = 10 - 15\,\text{mm of WC}$$

$$\text{Hepa filter} = 35 - 40\,\text{mm of WC}$$

Cooling coil = 2 rows deep = 2–4 mm of WC, 4 rows deep = 4–6 mm of WC and 6 rows deep = 7–10 mm of WC.

External static pressure (ESP)

External static pressure is very important for designing of duct.

1. For static pressure calculation in standard equipment such as fittings

$$\Delta P = C0 \times (V \div 4005)^{\wedge}2$$

Table 7 Static pressure in straight duct

Section	Dehumidified (CFM)	Velocity (FPM)	Friction (In wc/wft)	Length of duct (ft)	Static pressure (F*L/100)
P1-P2	9572	1385	0.065	20.97	0.013
P2-P4	6406	1260	0.065	23.35	0.015
P4-P5	3202	1065.2	0.065	21.43	0.013
P5-P7	1601	899.2	0.065	26.28	0.017
S1-S2	8309	1339.4	0.065	33.60	0.021
S2-S3	5473	1211	0.065	8.091	0.0052
S3-S4	3916	1116.6	0.065	19.15	0.012
S4-S5	2070	955.7	0.065	31.898	0.020
Total ESP in straight duct = 0.1162 inch of wc					

where $C0$ = coefficient of friction.

V = face velocity in fpm.

ΔP = static pressure in inches of WC.

For static pressure calculation in straight duct, consider largest path. Table 7 represents the static pressure in straight duct.

4 Drafting

HVAC drafting [2] is the engineering design which prepares using AutoCAD software. As deal with the HVAC service, it is called as the HVAC drafting. It includes the following details given below.

1. Machine layout (chillers, AHU, FCU)
2. Air terminal layout (diffuser, grills)
3. Ducting layout
4. Chilled water pipe layout
5. Refrigerant pipe layout
6. Insulation layout

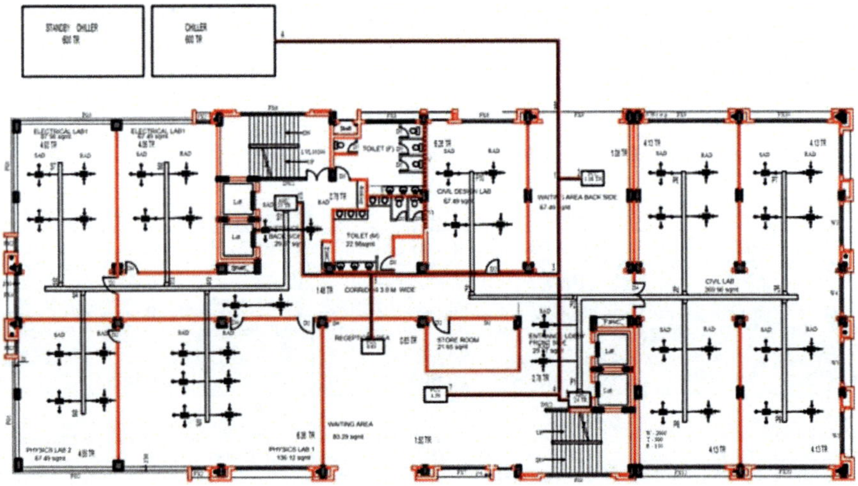

Fig. 2 HVAC layout of the building

7. Installation detail drawing

Figure 2 given below represents the HVAC layout of the ground floor of the building of educational institute. In this paper, we are unable to show layout of each floor. Since building has typical drawing layout, probably each floor has same dimension and size. So probably each floor will be the same HVAC layout.

5 Energy Conservation Strategy

Energy conservation [7] and saving are very important for HVAC system. HVAC strategy plays important role in energy efficient technology, while improving the safety of the building.

I. Properly insulate the cooled space.
II. Minimize the use of appliances and lighting.
III. For better efficiency, proper maintenance should be done at suitable time.
IV. Replace incandescent and CFL light with LEDs.
V. Use smart thermostat for saving energy.
VI. Use heat blocking curtains that stop the sun heat.
VII. Reduce hot water energy waste in the building.
VIII. Seal and insulate cooling or heating ducts increase the efficiency of cooling & heating system.

6 Energy Saving from the Building

VRF system is used for the top floor of the building where solar panel is established as power source for running the VRF system. Due to use of solar power, there is an energy saving which results in saving in annual cost- electrical consumption of 60 Ton VRV air conditioning contain = 40 KW.

Actual unit of electricity consumed per day = 80% of 8 h * 40 KW = 384 units electricity rate per unit in Greater Noida = Rs 7.

Electricity consumed per day in price = (7 * 384) = Rs 2688.

Electricity consumed per year in price = (2688 * 365) = Rs 98,112.

7 Discussion of Design

1. The given building with multi floors is designed in such a way that the ground floor to the 8th floor of the building operates with water cooled centrifugal central plant. Plant has been located on the back side of the building.

2. The top floor, i.e., 9th floor of the building is operated with the VRF system. The power source is obtained from the solar panel. The solar panel has been located on the roof of the building. This system has benefits in the form of energy savings.
3. Floor is divided into four zones, and each zone have air handling and fan coil unit as per the requirement.
4. Chiller plant has used primary and secondary pumping system for better functioning and operation and also save the energy.
5. Secondary pumps are inbuilt with pump logic control panel with variable frequency drive to run the pumps based on demand.
6. In pipe design, the concept of riser and floor piping has been used. Riser is for floor-to-floor piping, and floor piping is directly connected with FCU and AHU.
7. System is economically designed, easy in maintenance, and user friendly in operation.

8 Conclusions

Surely HVAC system has become a necessity for human; this paper elaborates the design of a centralized chilled water air conditioning system for the educational institute building as per the standard of ASHRAE and ISHRAE.

1. For proposed educational institute building, a central AC system used R-1233zd (E) refrigerant.
2. The total approximate load requires for educational building is 650 TR, and actual load is 633 TR.
3. Chillers have capacity of 600 TR designed for the central air conditioning system. There are two chillers required for the proposed system one is working chiller and other is stand-by chiller. Water cooled centrifugal chiller has been used for the better efficiency and easy operation.
4. The cost of the whole project is divided into two parts. First for chilled water system, the total cost for chilled water system is about 100 k to 110 k per ton. Second for VRV systems which have total cost is about 55 k to 60 k per ton. Hence, the total cost of the project will be Rs 6–7 crore approx.

Recommendation

1. There is a recommendation for ensuring energy saving use air curtains and air locks at the doors, windows to block the heating from the sun. Insulate the walls with insulating material pads.
2. Another recommendation is to use additive liquid naphthenic oil base contains two anti-wears, anti-oxidation systems which added to the refrigerants, and it results in about 5–20% reduction in the energy consumption.

Acknowledgements The authors are thankful to mechanical department Ganeshi Lal Bajaj institute of technology and management Greater Noida for providing better facilities for doing this research project.

References

1. Charyl, K.V., Mohiuddin, M.D.S., Khan, R.A.: Design of Central Air Conditioning System for a Multy-Storeyed Office Building (2014)
2. Ratna Kumari, K., Reddy, A.R., Sagar, M.V.: Design and Drafting of HVAC, Central Air Conditioning System for an Office Building (2016)
3. Zhang, Z., Zhang, Y., Khan, A.: Thermal comfort of people in a super high-rise building with central air-conditioning system in the hot-humid area of China. Energy Build (2019)
4. Arya, J.S., Chavda, N.K.: Design and Performance Analysis of Water Chiller (2014)
5. Park, D.Y., Chang, S.: Effects of combined central air conditioning diffusers and window-integrated ventilation system on indoor air quality and thermal comfort in an office. Sustain. Cities Soc. (2020)
6. Sahu, S.K.: Cooling Load Estimation for a Multi-story office building, Department of mechanical engineering, NIT Rourkela (2014)
7. Arora, Domakundwar: Refrigeration and Air-conditioning, 4th edn, Dhanpat rai, (2005)
8. Holman, J.P.: Air Conditioning Engineering, 5th edn. Tata McGraw Hill Book
9. The University of North Carolina of Chapel Hill, Department of facilities planning & construction, design guidelines (2010)
10. Khakre, V.V., Wankhade, A.M.: Review paper on Design and Computational Analysis of Air Flow through Cooling Duct (2014)

Competency of Alcoholic Fuels as Diesel Blends

Payal Sharma, Nathi Ram Chauhan, and Manish Saraswat

1 Introduction

Diesel engines overpower transportation sector on account of better thermal efficiency and robustness. On the contrary, they also majorly contribute to the air pollution as they emit unburnt hydrocarbons, carbon monoxide, oxides of nitrogen, soot, and smoke which are reported to be more hazardous than those from petrol engines. To fulfill the energy demands on economic grounds without harming the environment, we need promising alternative fuel with enhanced fuel properties [1]. Biodiesel transesterified from vegetable oils have been found favorable because of better quality exhaust emissions and analogous engine performance without any modification in diesel engines. Competency of alcoholic fuels has also been prioritized by many researchers as diesel blends for their substantial reduction in the exhaust emissions [2]. Alcoholic fuels are also found compatible with gasoline and blends [3]. Methanol and ethanol are the most promising alternatives for spark ignition engines since their combustion properties bear resemblance to gasoline [4]. Octane number of alcoholic fuels is higher than other conventional fossil fuels; concurrently they have low cetane number [5, 6]. Gasoline fuels can benefit from it as they enhance their octane rating, but for their usage in diesel engines, cetane enhancers like peroxides, nitrate, esters, azides, and ethers have to be employed [7]. Alcoholic fuels probe some struggle during ignition of air–fuel mixture in diesel engines and found to have longer ignition delay periods [8]. For this reason, the blending of alcoholic fuels with diesel needs careful consideration and various methods are employed for that like emulsion, dual injection, fumigation, etc.

P. Sharma (✉) · N. R. Chauhan · M. Saraswat
Department of Mechanical and Automation Engineering, Indira Gandhi Delhi Technical University for Women, New Delhi, India

M. Saraswat
Department of Mechanical Engineering, Lloyd Institute of Engineering and Technology, Greater Noida, India

L. M. Das et al. (eds.), *Recent Trends in Thermal Engineering*,
Lecture Notes in Mechanical Engineering,
https://doi.org/10.1007/978-981-16-3428-4_13

As conceived from the literature, alcohol–diesel blends tend to have long ignition delay periods, and this period increases as the dosage of alcohol is increased in the blend [9]. Conventional oxygenated fuels such as tert-Amyl methyl ether and methyl tert-butyl ether were not fit with diesel although alcoholic fuels also found to have some miscibility issues. Phase separation is observed in ethanol–diesel blends at low temperatures, but ethanol in diesel helps minimize CO, NO_X emissions, and particulate matter. Similarly homogeneity of methanol–diesel blends is also troublesome, but it keeps seeking the attention of researchers because of its low cost and high oxygen fraction [10, 11]. Butanol on the other hand dominates over ethanol and methanol as diesel blend as it forms a better blend when mixed in diesel with no phase separation [12, 13]. Also cetane number of butanol is high which makes it a favorable additive for diesel blends. Cetane number is an important specification indicating the self-ignitability of the fuel. Ignition delay period can be reduced on addition of cetane enhancers. Literature studies have revealed the impact of cetane number of oxygenated fuels on engine emissions [14]. With increase in cetane number, nitrogen oxide, carbon monoxide, and sulfur dioxide emissions have seen to be reduced with a little increase in soot. Alcoholic additives play along with cetane enhancers to boost the ignition characteristics and operation of diesel engine [15].

2 Comparing Different Alcoholic Fuels as Additives

If miscibility is to be discussed, methanol and ethanol have limited solubility in diesel fuel, but stable blends can be formulated by adding certain additives. Higher alcohols do not face such problems because of better solubility with diesel, even they form stable blends at low temperature. Methanol and ethanol possess low flash point than diesel fuel, and butanol having the long carbon chain than primary alcohols has high flash point because of which it is safe to handle and less corrosive than primary alcohols [16]. Dogan (2011) reported the impact of n-butanol blends in diesel at different volume percentages varying from 5 to 20%. With increase of volume percentage of n-butanol in diesel, NO_X, CO emissions, and smoke were reported to be reduced but with increase in unburnt hydrocarbon. An increase was found in the BSFC and in BTE accompanied by decrease in exhaust gas temperature with increase in n-butanol dosage in diesel [17].

Karabektas and Hosoz (2009) explored isobutanol–diesel blends with varying volume percentage of isobutanol from 5 to 20% and compared with neat diesel. Reduction in CO and NO_X emissions was observed with increment in isobutanol content in diesel again with corresponding increase of HC emissions. BSFC increased proportional to increase in isobutanol percentage in diesel. As for BTE, neat diesel was observed having the largest BTE, but the blend containing 10% isobutanol showed a little enhancement in BTE at high level speeds [18].

Many researchers have also focused on the approach considering multicomponent fuel blends. Biodiesel which is also an encouraging substitute for diesel engines can also be used as an emulsifier to advance the solubility of ethanol with diesel. Chang et al. (2013) integrated diesel blend with acetone–butanol–ethanol (ABE) solution by 20% and 0.5% volume of water and found an improvement in BTE from regular diesel and also found reduction in NO_X emissions, PM, and unburnt HCs. The blended fuel also passed the stability test and was stable even with water concentration of 0.5–1% [19].

Bilgin et al. (2002) experimentally evaluated ethanol–diesel blends for 2, 4, and 6% of ethanol in diesel on variable compression ratio engine to determine the optimal compression ratio for best performance and efficiency of this blend. Better results were obtained for 4% ethanol in diesel blend as brake force, brake power, and BTE was increased alongside decrease in BSFC in contrast to when operated on diesel alone. At compression ratio of 21, great efficiency is said to be reported [20]. Rakopoulus et al. (2008) experimentally investigated ethanol blending with diesel in 5 and 10% by volume with engine at two speeds and three loads. CO and NO_X emissions were little minimized at higher percentage of ethanol in blend with an increase in unburnt hydrocarbons at higher percentages of ethanol in diesel blend as compared to regular diesel. With raise in percentage of ethanol in blend, BSFC was noticed to increase with slight increase in BTE [21].

As we discussed, miscibility of ethanol with diesel is troublesome at low temperatures. Biodiesel improves the solubility of ethanol in diesel, but again because of its poor cold flow characteristics, biodiesel possesses some problems with diesel in much cold weathers. Mofijur et al. (2015) reviewed ethanol–biodiesel–diesel blends and summarized that the observed trends show remarkable reduction in NO_X, PM, HC, and smoke emissions with little increase in BSFC [22]. However, many researchers suggested emulsifiers other than biodiesel for ethanol and diesel blends. Shi et al. (2005) prepared ethanol–methyl soyate blends and added to diesel at 15 and 20% volume percentage. Another blend of diesel and 20% methyl soyate was also prepared to compare. Methyl soyate contains less oxygen content than ethanol. Reduced PM emissions were observed with increasing oxygen content in fuel. So 20% ethanol–methyl soyate blend in diesel produced less PM emission but highest NO_X among all samples. Diesel with 20% methyl soyate produced less HC emission as compared to regular diesel, whereas HC emissions were higher for ethanol–methyl soyate–diesel blends [23].

Huang et al. (2009) suggested n-butanol as an additive for ethanol–diesel blends. They conducted experimental investigation on ethanol–diesel blends with 10, 20, 25, and 30% of ethanol in diesel without any supplement and found the blends separating in two different layers after some time. To solve the miscibility and stability issues, they added 5% n-butanol as supplement to all the prepared samples of ethanol–diesel blends, which kept the blends stable for not less than 11 days. However, percentage of n-butanol can be optimized, but the study revealed n-butanol to be good emulsifier to mix ethanol with diesel. With increasing percentage of ethanol, fuel consumption showed an increasing trend [24]. Xingcai et al. (2004) experimentally evaluated ethanol with diesel blends at 10 and 15% ethanol blended with diesel for emission and combustion characteristics and made comparison with base diesel fuel. Cetane enhancers were also used to test their effect on emissions. BSFC increased with increasing ethanol content at overall operating conditions. For 10% and 15% ethanol–diesel blend, BTE increased by 1–2.3%. CO, NO_X emissions, and smoke minimized with increase in HC emissions on increase in ethanol content. Cetane enhancer positively impacted CO and NO_X emissions but was not favorable for HC emissions [25].

As discussed, poor cold flow characteristics, high viscosity, and low volatility makes biodiesel unsuitable with diesel in cold weather conditions. Methanol is added to biodiesel–diesel blends to upgrade the characteristics at low temperature and also increases the oxygen content which can notably minimize PM emissions. Qi et al. (2010) investigated methanol as additive to biodiesel–diesel blends in volume percentage of 5 and 10% added (BDM5 and BDM10, respectively) to blend of 50% biodiesel–50% diesel (BD50). At low engine loads, a delay in combustion was observed for BDM5 and BDM10 in contrast to BD50. At low speeds, CO emissions for BDM5 and BDM10 were lower than BD50, whereas HC and NO_X emissions were very much similar [26]. MatYasin et al. (2014) also used methanol to enhance the viscidity and density of biodiesel and diesel blend. Performance was tested for 5% by volume of methanol in 20% biodiesel–diesel blend (B20M5) and B20 at different engine speeds and compared with base diesel. With rise in engine speed, BSFC for base diesel decreases, whereas BSFC for B20 and B20M5 increased. CO emissions significantly reduced for B20 and B20M5, but NO_X was high as compared to base diesel [27].

Now another issue of blending methanol to biodiesel–diesel blend is reduction in cetane number of blend which would affect the self-ignitability of fuel. Li et al. (2014) investigated and suggested some cetane enhancers for methanol in biodiesel–diesel blends to enhance the combustion properties of fuel. Results revealed few cetane enhancers significantly reducing emissions especially cyclohexyl nitrate which was claimed to minimize ignition delay period, hence improving the ignition characteristics [28].

Isopropanol having high cetane number and better miscibility in diesel makes it better option as compared to methanol and ethanol. Alptekin (2017) compared ethanol and isopropanol blends in diesel and concluded with moreover similar results, except there were no solubility issues with isopropanol–diesel blends at room temperature [29]. However, solubility issues of ethanol–diesel blends can be resolved by some emulsifiers as discussed previously in this paper. Atmanli (2016) made comparative research investigation on diesel–waste oil with 20% alcohol (n-butanol, propanol, and 1-pentanol) one by one taken as three samples. It was concluded that higher alcohols are more powerful in NOx reduction and claimed to enhance the cloud point [30] (Table 1).

Table 1 Effect of addition of alcoholic additives to diesel

Addition of alcoholic fuels to diesel	Cetane number of blend	Emissions	Properties	Barriers
1. Methanol (CH_3OH)	Decreases	Notable reduction in smoke and CO emission, reduction in NO_x emissions with increase in CO and HC emissions as content of methanol in blended fuel is increased	Immiscible in diesel fuel, higher oxygen fraction, low sulfur content	Low flash point is a drawback for safety, low ignitability
2. Ethanol (C_2H_5OH)	Decreases	Substantial reductions in particulate matter (PM), significantly reducing SO_2 emissions, NO_x and CO_2 emissions of blends are minimized somewhat; even total HC emissions are less than that of diesel fuel	Viscosity and lubricity of blend decreases, limited miscibility in diesel fuel	Poor fuel economy, low ignitability, causes ignition delay, low calorific value, extremely low cetane number
3. Propanol (C_3H_7OH)	Decreases	Decreased NOx at low load but minutely increased at high load, increase in BSFC because of low heating value of propanol as compared to diesel, smoke opacity minimized	Better solubility in diesel fuel, better miscibility, highest heating value and highest cetane number among primary alcohols	–
4. n-Butanol (C_4H_9OH)	Decreases	Considerable drop in soot, smoke density, reduced CO, NO_x, and PM emissions, rise in BSFC and BTE and reduced exhaust gas temperature with rise in n-butanol dose in fuel blends with respect to that of reference diesel fuel	Can be mixed with diesel without phase partition, greater miscibility than ethanol	–
5. Isobutanol (C_4H_9OH)	Decreases	Reduction in CO and NO_x emissions, increase in HC emission lead to reduced BTE proportional to isobutanol dosage, increase in BSFC	Stability with diesel fuel for long duration with no phase partition, specific gravity, and latent heat of vaporization of isobutanol are almost adjacent to the properties of diesel	Low cetane number of isobutanol can worsen auto ignition capability of injected fuel

3 Future Scope

Proper blended ratio of alcohol and diesel can be used as a base fuel for addition of other additives so as to cover up the anomalies of primary alcohols and to achieve better fuel properties. The ideal percentage for blending ratio of alcohols with diesel is the point to probe to achieve desired reduction in emission, increase in thermal efficiency, and for other factors like ignition delay, peak cylinder temperature, etc. [31]. Between the ranges of these alcohol series, selection of suitable alcoholic additive is important along with optimal blending ratio [32].

4 Conclusion

As conceived from the literature, primary alcohols possesses troublesome behavior and require cetane enhancers and certain additives to solve miscibility and stability issues. Although, cetane enhancers are always required while mixing alcoholic additives with diesel because of their low cetane numbers. Methanol and ethanol having the lowest cetane number amongst other alcoholic fuels have miscibility issues and find difficulty in forming a stable mixture. Many researches are focused and can be further probed to search additives and emulsifiers for stable mixtures of diesel with primary alcohols. Butanol on the other hand has some superior properties among other alcoholic fuels and can be further probed to find proper blends to reduce the hydrocarbon emission which is still increasing in butanol-diesel blends.

References

1. Babu, D., Anand, R.: Influence of fuel injection timing and nozzle opening pressure on a CRDI-assisted diesel engine fueled with biodiesel-diesel alcohol fuel. In: Advances in Eco-Fuels for a Sustainable Environment 08-102728-8.00013-9 (2019)
2. Yasar, A., Bilgili, M., Yildizhan, S.: The influence of diesel-biodiesel-alcohol blends on the performance and emissions in a diesel engine. Int. J. Sci. Technol. Res. **1**(7), 2422–8702 (2015)
3. Saraswat, M., Chauhan, N.R.: Assessment study of butanol-gasoline blends in variable compression ratio spark ignition engine. J. Sci. Ind. Res. **77**, 405–409 (2018)
4. Siwale, L., Kristof, L., Adam, T., Bereczky, A., Mbarawa, M., Penninger, A., Kolesnikov, A.: Combustion and emission characteristic of blends: -n-butanol-diesel (D2); and dual alcohols: n-butanol-methanol with gasoline in internal combustion engines. Dev. Combust. Technol. (2016). https://doi.org/10.5772/64747
5. Allawi, M.K.: Effect of alcohol additives in gasoline engine performance and emissions. Int. J. Adv. Res. **4**, 2320–9135 (2016)
6. Malhotra, R.K., Das, L.M.: Biofuels as blending components for motor gasoline and diesel fuels. J. Sci. Ind. Res. **62**, 90–96 (2003)

7. Cataluna, R., Silva, R.D.: Effect of cetane number on specific fuel consumption and particulate matter and unburned hydrocarbon emissions from diesel engine. J. Combust. (2012). https://doi.org/10.1155/2012/738940
8. Ali, O.M., Abdullah, N.R., Mammat, R., Abdullah, A.A.: Comparison of the effect of different alcohol additives with blended fuel on cyclic variation in diesel engine. Energy Procedia **75**, 2357–2362 (2015)
9. Ribeiro, N.M., Pinto, A.C., Quintella, C.M., Rocha, G.O.D., Teixeira, L.S.G., Guarieiro, L.L. N., Rangel, M.D.C., Veloso, M.C.C., Rezende, M.J.C., Cruz, R.S.D., Oliveira, A.M.D., Torres, E.A., Andrade, J.B.D.: The role of additives for diesel and diesel blended (ethanol or biodiesel) fuels: a review. Energy Fuels **21**, 2433–2445 (2007)
10. Zhang, C., Zhai, X., Li, Y., Sun, Z.: Research on combustion characteristics and emissions of methanol-diesel fuel with different additives. Adv. Mater. Res. **354–355**, 462–467 (2012)
11. Soni, D.K., Gupta, R.: Application of nano emulsion method in a methanol powered diesel engine. Energy **126**, 638–648 (2017)
12. Saraswat, M., Chauhan, N.R.: Comparative assessment of butanol and algae oil as alternate fuel for SI engines. Eng. Sci. Technol. Int. J. **23**, 92–100 (2020)
13. Chotwichien, A., Luengnaruemitchai, A., In, S.J.: Utilization of palm oil alkyl esters as an additive in ethanol–diesel and butanol–diesel blends. Fuel **88**, 1618–1624 (2009)
14. Niculescu, R., Clenci, A., Simen, V.I.: Review on the use of diesel–biodiesel–alcohol blends in compression ignition engines. Energies **12**, 1194 (2019)
15. Fayyazbakhsh, A., Pirouzfar, V.: Comprehensive overview on diesel additives to reduce emissions, enhance fuel properties and improve engine performance. Renew. Sustain. Energy Rev. **74**, 891–901 (2017)
16. Bannikov, M., Gillani, S.E., Vasilev, I.: Effect of alcohol additives on diesel engine performance and emissions. Mater. Methods Technol. **9**, 1314–7269 (2015)
17. Dogan, O.: The influence of n-butanol/diesel fuel blends utilization on a small diesel engine performance and emissions. Fuel **90**, 2467–2472 (2011)
18. Karabektas, M., Hosoz, M.: Performance and emission characteristics of a diesel engine using isobutanol–diesel fuel blends. Renew. Energy **34**, 1554–1559 (2009)
19. Chang, Y.C., Lee, W.J., Lin, S.L., Wang, L.C.: Green energy: water-containing acetone–butanol–ethanol diesel blends fueled in diesel engines. Appl. Energy **109**, 182–191 (2013)
20. Bilgin, A., Durgun, O., Sahin, Z.: The effects of diesel-ethanol blends on diesel engine performance. Energy Sour. **24**, 431–440 (2002)
21. Rakopoulos, D.C., Rakopoulos, C.D., Kakaras, E.C., Giakoumis, E.G.: Effects of ethanol–diesel fuel blends on the performance and exhaust emissions of heavy duty DI diesel engine. Energy Convers. Manage. **49**, 3155–3162 (2008)
22. Mofijur, M., Rasul, M.G., Hyde, J.: Recent developments on internal combustion engine performance and emissions fuelled with biodiesel-diesel-ethanol blends. Procedia Eng. **105**, 658–664 (2015)
23. Shi, X., Pang, X., Mu, Y., He, H., Shuai, S., Wanj, J., Chen, H., Li, R.: Emission reduction potential of using ethanol–biodiesel–diesel fuel blend on a heavy-duty diesel engine. Atmos. Environ. **40**, 2567–2574 (2006)
24. Huang, J., Wang, Y., Li, S., Roskilly, A.P., Yu, H., Li, H.: Experimental investigation on the performance and emissions of a diesel engine fuelled with ethanol–diesel blends. Appl. Therm. Eng. **29**, 2484–2490 (2009)
25. Xingcai, L., Zhen, H., Wugao, Z., Degang, L.: The influence of ethanol additives on the performance and combustion characteristics of diesel engines. Combust. Sci. Technol. **176**, 1309–1329 (2004)
26. Qi, D.H., Chen, H., Geng, L.M., Bian, Y.Z.H., Ren, X.C.H.: Performance and combustion characteristics of biodiesel–diesel–methanol blend fuelled engine. Appl. Energy **87**, 1679–1686 (2010)
27. MatYasin, M.H., Yusuf, T., Mamat, R., Yusop, A.F.: Characterization of a diesel engine operating with a small proportion of methanol as a fuel additive in biodiesel blend. Appl. Energy **114**, 865–873 (2014)

28. Ruina, L., Zhong, W., Peiyong, N., Yang, Z., Mingdi, L., Lilin, L.: Effects of cetane number improvers on the performance of diesel engine fuelled with methanol/biodiesel blend. Fuel **128**, 180–187 (2014)
29. Alptekin, E.: Evaluation of ethanol and isopropanol as additives with diesel fuel in a CRDI diesel engine. Fuel **205**, 161–172 (2017)
30. Atmanli, A.: Comparative analyses of diesel–waste oil biodiesel and propanol, n-butanol or 1-pentanol blends in a diesel engine. Fuel **176**, 209–215 (2016)
31. Balamurugan, T., Nalini, R.: Experimental investigation on performance, combustion and emission characteristics of four stroke diesel engine using diesel blended with alcohol as fuel. Energy **78**, 356–363 (2014)
32. Chen, H., Zhou, Z., He, J., Zhang, P., Zhao, X.: Effect of isopropanol and n-pentanol addition in diesel on the combustion and emission of a common rail diesel engine under pilot plus main injection strategy. Energy Rep. **6**, 1734–1747 (2020)

Experimental Study of Performance and Emission Characteristics Using Biodiesel Blends in a Diesel Engine: A Review

Rahul Sharma, Roshan Raman, Omprakash Yadav, Vimal Pratap Singh, Himanshu Tanwar, Uday Kumar, Udit Gupta, and Rahul Shakye

Abbreviations

ATDC	After top dead center
BSCO	Brake-specific carbon monoxide emission
BSEC	Brake-specific energy consumption
BSFC	Brake-specific fuel consumption
BSFL oil	Black soldier fly larvae oil
BSHC	Brake-specific hydrocarbon emission
$BSNO_x$	Brake-specific oxides of nitrogen (g/kWh)
BTDC	Before top dead center
BTE	Brake thermal efficiency
CAD	Crank angle degrees
CeO_2	Cerium oxide
CFB	Cymbopogon flexuosus biofuel
CHP	Combined heat and power
CI	Compression ignition
CNSL	Cashew nut shell liquid biofuel
CNT	Carbon nanotubes
CO	Carbon monoxide
CO_2	Carbon dioxide
CRDI	Common rail direct injection

R. Sharma · O. Yadav
Assistant Professor, Department of Mechanical Engineering, G.L. Bajaj Institute of Technology and Management, Greater Noida, UP 201308, India

R. Raman
Assistant Professor, Department of Mechanical engineering, The Northcap University, Gurugram 122017, India

V. P. Singh (✉) · H. Tanwar · U. Kumar · U. Gupta · R. Shakye
B.Tech. Student, Department of Mechanical Engineering, G.L. Bajaj Institute of Technology and Management, Greater Noida, UP 201308, India

L. M. Das et al. (eds.), *Recent Trends in Thermal Engineering*,
Lecture Notes in Mechanical Engineering,
https://doi.org/10.1007/978-981-16-3428-4_14

CV	Calorific value
DF	Dual fuel
DI	Direct injection
DPME	Date pit methyl ester
EBD	Ethanol–biodiesel–diesel blend
ECU	Engine control unit
EEFFAO	Ethyl esters of free fatty acid oils
EGR	Exhaust gas recirculation
EGT	Exhaust gas temperature
FD	Fossil diesel
GHG	Greenhouse gas
GHG	Greenhouse gases
HC	Unburnt hydrocarbon
HCCI	Homogeneous charge compression ignition
HFRR	High-frequency reciprocating rig
HLB	Hydrophilic–lipophilic balance
HRR	Heat release rate
IC	Internal combustion
ID	Ignition delay
IDI	Indirect injection
ISB	Isobutanol
ISFC	Indicated specific fuel consumption
ITE	Indicated thermal efficiency
LGBD	Lemon grass biodiesel
LGO	Lemongrass oil
LHRE	Low heat rejection diesel engine
LPO	Lemon peel oil
LTC	Low-temperature compression
MEMSO	Methyl ester of mango seed oil
NO_x	Nitrogen oxide
PCCI	Premixed-charge compression ignition
PEN	Pentanol
PM	Particulate matter
PPLTC	Partially premixed low-temperature combustion
RME	Rapeseed methyl ester
RO	Rapeseed oil
SF	Surfactant
SFC	Specific fuel consumption
SME	Soybean methyl ester
SOI	Start of injection
VCR	Vapor compression ratio

1 Introduction

The diesel-fueled engine is generally used worldwide. The consumption of petroleum fuel resources and atmospheric air pollution is the major issue in today's world. This problem can be solved by alternative energy resources and biofuel looks promising as an alternative energy resource. It is utilized in diesel engine as an elective fuel by making its mixes with diesel. Many types of biofuels have been tested on diesel engines, including sunflower oil, Pistacia lentiscus, safflower oil, macroalgae, jatropha, pongamia, linseed, pine oil, and many others that gives the close BTE as of diesel and the emission got reduced in considerable amount. In the ongoing decade, the primary goal is to produce biodiesel from edible oils such as cottonseed oil, sunflower oil, and coconut oil. Biodiesel made from edible oils can have a detrimental effect on farming due to a lack of food crops, so non-consumable oils are used for biodiesel production. The primary advantages of using biodiesel are its transportability, availability, improved burning efficiency, lower sulfur content, and higher cetane number, better lubricity property, increased biodegradability, domestic origin, increased lubrication, and higher flash point. They contain unalienable oxygen, which degrades hydrocarbon and sulfur content in the same way as gases do. Engine experts have observed that by utilizing biodiesel, (NO_x) emission rate intensifies while carbon monoxide (CO), hydrocarbon (HC), and particulate matter (PM) decline in contrast with diesel fuel [1, 2]. In this paper, we have reviewed some biofuel research papers in the related to performance and emissions. This paper gives a generalized idea about the performance and emanation variation using biofuel with respect to diesel.

1.1 Preparation of Biofuels

The biofuels or cycle stage are gathered into three ages. The biofuels that are procured from the, beats or food crops, juice of sugar sticks are ethanol and biodiesel. Bioethanol and biodiesel are made from food that is unfit for human use. However, land inadequacy has become a problem with second-generation non-edible yields for food improvement. Furthermore, third-generation biofuels are made from cyanobacteria, microalgae, and other microorganisms, which may be the most reassuring way to treat and control the industry which in future may gratify the overall energy requirement (Fig. 1).

Bioethanol from grain and cellulosic, biofeed for CHP turbines and gas turbines, torrefaction of biomass, biopyrolysis, and biodiesel are the recent developments on biofuels with their genuine variables. Biodiesel production is now in the growing stage, according to a lifecycle development assessment, and other advancements are in the rising stage as well. However, transesterification of plant vegetable oils, animal fats, and waste cooking oil yields this fuel. Moreover, various tests on diesel engines using crude vegetable oil were conducted prior to the development of

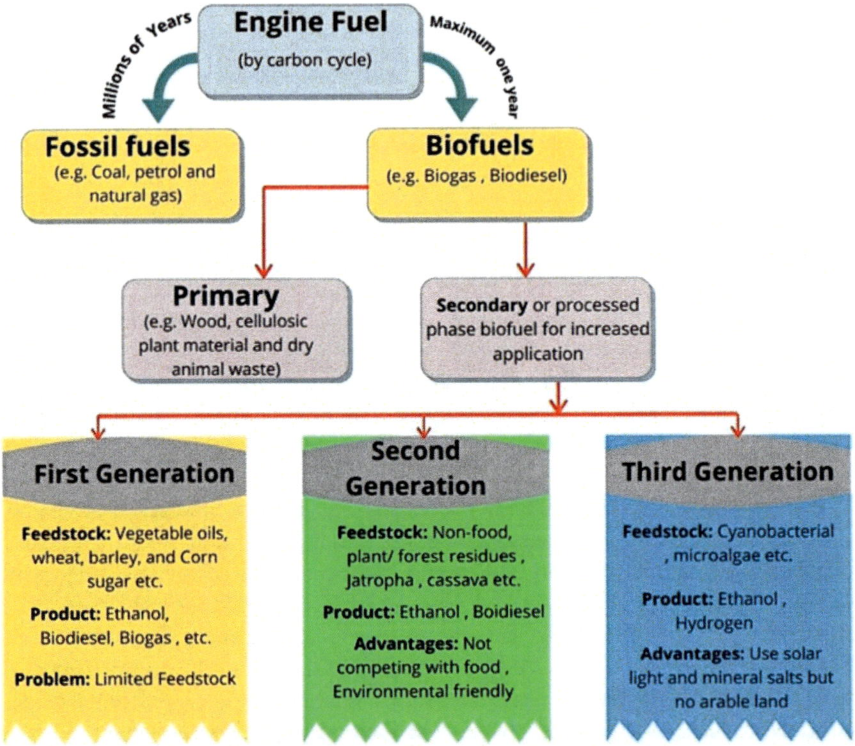

Fig. 1 Development stages of biofuel

biodiesel. It was expansion in emanations as vegetable oil having high consistency, low instability besides, their polyunsaturated character. Through the transesterification cycle kinematic consistency augmented and its calorific worth was diminished.

2 Biofuel as an Alternative Fuel in Diesel Engine

The performance and emission parameters of biodiesel-fueled diesel engines were investigated by Puneet Verma et al. He discovered that the main challenge with using vegetable-based biofuel in diesel engines was higher viscosity, which can be overcome by the transesterification process. However, the majority of biodiesel contributes to improved brake power and reduced BSFC. The results of exhalations showed that NO_x is increased most of the time, while HC, CO, and PM deposits are reduced at substantial level. The perfect biodiesel-to-diesel mixture for the engine was discovered to be a B20 blend. Hence, biodiesel is a suggested to be utilized as alternative fuel in conventional diesel engines with environmental benefits [1].

According to Dhinesh et al., because of the accessibility of oxygen atoms in the biofuel, CFB and its blends caused lower brake thermal performance and increased BSFC, as well as lower HC, CO, and smoke emissions when compared to diesel fuel, though they caused higher rates of NO_x and CO_2 [2]. Mofijur et al. stated that owing to the higher oxygen content in biofuels, both biodiesels–diesel and ethanol–biodiesel–diesel blending play a significant role in reducing exhaust gas emissions, such as CO, HC, and PM. However, because of the higher cetane number, higher oxygen content, injection timing, and other characteristics of biodiesel and ethanol, ethanol–biodiesel–diesel and biodiesel–diesel blends emit more NO_x [3]. Rajesh Kumar and colleagues investigated the effect of dimethyl carbonate (DMC) in a diesel engine at partially premixed low temperature condition. It was observed that DMC15 performed better with least smoke and NO_x levels while the infusion timing was changed from 21° to 25° CA bTDC. By and by DMC15 infused at 21° CA bTDC under 30% EGR experienced longest start delay and most minimal tops in weight and Heat release rate (HRR) to accomplish a greatest concurrent decrease of NO_x outflows (46.1%) and smoke haziness (64.7%). Both hydrocarbons (THC) and carbon monoxide (CO) outflows expanded at increasing value of EGR rates [4]. According to Khiari et al. analysis Pistacia lentiscus (PL) biodiesel has a 3% higher thermal efficiency than diesel fuel [5]. Raju, Balasubramanian, and colleagues investigated the performance and emission characteristics of a diesel engine using low reactivity fuel and biodiesel-blended nanoparticles. Different blend proportions were used: B20-80, B20A25, and B20A50. The findings suggest that in-cylinder pressure peaks have a significant impact on engine performance. Moreover, when using proper mixture, NO_x and smoke emissions are significantly reduced, with a small increase in HC and CO, but when using nanoparticles, exhaust emissions are further reduced [6]. Vara Prasad and colleagues investigated the performance of a four-stroke diesel engine using coconut oil as a biofuel. The results of the engine tests show that using diesel coconut oil blends reduces fuel consumption significantly. Furthermore, the thermal efficiency of the engine is improved significantly when biodiesel blends are used [7]. Leite, Santos, and colleagues investigated how soybean, linseed, and crambe biodiesel affected the performance and emission parameters of a diesel engine. The fuels were tested in a 5-kVA generator engine with a power range of 750–3000 W [8] (Table 1).

3 Summary

Based on the findings of the above-mentioned literature survey, it can be concluded that biofuel, in the form of biodiesel blends, can be easily used in diesel engines without requiring any engine modifications. The parameters of engines performance and emissions in generalized form can be summarized in the following points [1–44]:

Table 1 Summary of experimental investigation by various researchers

Author	Engine used	Performance	Emission	Combustion
Verma et al. [1]	Single cylinder, four-stroke diesel engines	Compared to petroleum diesel, there is increment in BSFC and abatement in BTE	Commendable reduction in HC and CO emanations NO_x emanations increase	–
Dhinesh et al. [2]	Single-cylinder diesel engine at 1500 rpm engine speed	BSFC is higher	Lower hydrocarbon (HC) emission and smoke emissions for the entire load range Lower CO and NO_x and increase in CO_2	Decrease in the cylinder pressure conditions as percentage of the CFB blend increases Lower HRR Fall in EGT
Huang et al. [9]	Four-cylinder diesel engine at 1800 rpm engine speed	BSFC increments with increment in % of EGR rate. BTE diminishes with increment in % of EGR rate	Under low load conditions soot emission decreased whereas as at high load conditions it slightly increased CO and THC emissions decreased. NO_x level rose up to 80% load. However, EGR reduces NO_x emissions Soot emissions increase slowly at low EGR rate then it increases sharply	Longer ignition delays. When EGR rate is low, peak in chamber pressure increases whereas with increment in EGR rate it decreases drastically
Mofijur et al. [3]	Single-cylinder diesel engine	–	Majority of the results shows that CO, THC, and PM decrease while CO_2 emission increases	–
Kumar et al. [4]	Four-stroke, diesel engine		THC and CO emissions increased at all EGR rates. Reduction in NO_x and smoke emission with EGR and injection timing of 21° CA bTDC	Longer ignition delay. With advanced in infusion timing from 25° to 21° CA bTDC, in-chamber pressure and HRR increments

(continued)

Table 1 (continued)

Author	Engine used	Performance	Emission	Combustion
Kumar et al. [10]	Four-stroke, diesel engine	BTE decreases with late and early SOIs. A slight reduction in efficiency during late and early SOIs	HC and CO outflows increment with expanding EGR rates. HC and CO emissions rise suddenly during the late injection. Smoke opacity increases as SOI is retarded and EGR is added. Decrease in NOx with late SOI and EGR	Longer ignition delay. As SOI retarded, HRR and in-cylinder pressure decreases but case is opposite with EGR
Annamalai et al. [11]	Single-cylinder cooled diesel engine	BSEC of all fuels used decreases with engine load which varies from no load to full load condition The BTE of all the fuels utilized increases as the load is increased	HC and NO_x emissions increases CO emissions decrease as load increases in the following order: neat diesel > LGO emulsion > LGO > LGO nano emulsion. HC emission decreases with addition of nanoparticle	Shorter ignition delay Decrease in cylinder peak pressure and HRR
Khiari et al. [5]	Air-cooled diesel engine	Reduction in BSFC. At high load, BTE increases	CO, HC, and particulates for the BD100 are slightly higher at low and moderate engine loads and slightly lower at high engine loads. NO_x emission increases for entire load range	With both oils, the pressure peak increases with engine load. At low and high engine load, HRR decreases. But BD100 exhibits higher HRR. Injection timing increases with increase in PL biodiesel fraction and engine loads
Vijayaraj et al. [12]	Kirloskar TAF 1 engine	BSFC increments with increment in biodiesel mixes. Neat biofuel shows higher BSFC than other fuel mixes. If the engine load and the amount of diesel in the blend increase, the BTE decreases	CO emissions are higher at low-to-medium engine loads. At full burden, carbon dioxide discharge of neat biofuel is extremely high as contrast with different mixes	The maximum pressure in the chamber and the rate of heat escape are also smaller Fall in exhaust gas temperature

(continued)

Table 1 (continued)

Author	Engine used	Performance	Emission	Combustion
			Decrease in HC emanations At full burden, NO_x release for B25 and B100 was 5.88% and 24% higher than that of diesel	
Raju et al. [6]	Single-cylinder diesel engine	Increment in Bsfc for entire load range BTE decreases at all loads, although the addition of a nanoparticle improves brake thermal efficiency somewhat at higher loads	HC and CO were high at low and part loads in case of different blends whereas with use of nanoparticles the HC and CO emissions reduce	For both dual fuel procedures, the in-chamber pressure and temperature are lowered, while the temperature is slightly higher on account of G-B20A50 Neat diesel fuel and G-B20A50 results in high HRR heat. While for other blends it is low
Han et al. [13]	Four-cylinder diesel engine	Higher specific fuel consumption is obtained	Low NO_x and smoke emission	Chamber pressure is lower. Because of its lower energy capacity, *n*-butanol needs a longer infusion time to achieve equal fuel energy
Piloto-Rodrígue et al. [14]	Four-stroke, four-cylinder diesel engine	Lower BTE and BSEC	Higher CO_2 Reduction in CO and NO_x	High combustion pressure rise rate Higher ignition delay High HR
Babu et al. [15]	Four-cylinder, DI CI engine	Increase in BSFC and BTE	With a higher butanol fraction, CO and UHC outflows increase. Pentanol use increases UHC and CO emissions, particularly at low and medium loads. Also, soot emissions are reduced	Results in delay in infusion and longer ignition delay with high EGR rate

(continued)

Table 1 (continued)

Author	Engine used	Performance	Emission	Combustion
Ashok et al. [16]	Air-cooled diesel engine	In all loading conditions, there is a reduction in BSFC. Increase in brake thermal efficiency BSEC is low	Lower CO emission and HC emission Higher NO_x emission. Lower smoke emissions at low and part load conditions	Ignition delay period became longer Higher in-chamber pressure
Prasad et al. [7]	Four-stroke single-cylinder diesel engine	Brake-specific fuel consumption and brake thermal efficiency are reduced	–	–
Srihari et al. [17]	Four-stroke, air-cooled, high-speed diesel engine	BSFC increases, particularly at part load conditions. At lower load conditions, the Diesel-E15 mode had a low BSFC	HC emissions and CO emissions increases. A most extreme increment is seen in E15D-Diesel mode In all modes when E15D is used as a fuel, higher CO emissions are observed Abatement in CO_2 and NO_x discharges	–
De Poures et al. [19]	Single-cylinder diesel engine	BSFC increments with an expansion in the amount of 1-Hexanol in the blend but it diminishes with EGR Decrement in the BTE is noted addition of l-hexanol in the mix and with EGR	CO discharge increments with and without EGR. The HC outflow increments with an expansion in the amount of l-Hexanol in the blend and it further expands with EGR The decrease in NO_x with and without EGR	The chamber pressure increments with load and decreased with an increment in proportion of l-Hexanol in the mix and for EGR Increase in combustion duration and it also increases further with use of EGR

(continued)

Table 1 (continued)

Author	Engine used	Performance	Emission	Combustion
Alagu et al. [21]	Kirloskar water-cooled compression ignition engine	Increase in BSFC for different mix of WHB–Diesel For an increase in biofuel concentration, BTE decreases	HC, CO, and smoke emissions reduce substantially However, the CO_2 and NO_x levels were impressively higher, particularly at part load conditions	Decrease in ignition delay at all engine load conditions All the WHB–Diesel blends exhibit lowered cylinder pressure, heat release rate EGT increases with increase in engine load
Hellier et al. [22]	1-cylinder diesel engine	HC and CO were escalated in case of DPME and RME. NOx are lowest in case of DPME due to lower adiabatic flame temp. and lower local temp PM is lower in all three blends but lowest in DPME	–	DPME shows similar ID as RME and SME. In cylinder global temp. was high for DPME HRR of DPME is higher than that of RME and SME
Kamarulzaman et al. [23]	Single-cylinder, four strokes, CI engine	BSFC increased in blended mode at all load conditions in comparison to neat diesel At low load, the BTE is lower. But at medium and full load, the 75% concentration of biofuel blend gives close results as that of diesel	CO, CO_2 and HC emanations are increased in greater extent Decrement in NO_x and O_2 emanations for all blends	ID is small for BSFL and improves when blended with diesel The peak chamber pressure and HRR are lower for BSFL by 3.28% and 13.38%, respectively EGT increases as compared to diesel in all load conditions
Deivajothi et al. [26]	Single-cylinder DI compression ignition engine	In case of all blends, the BSFC was higher throughout low to full load. At full load, B20 (20% fraction blend) shows closer results to that of diesel but still greater than that of diesel	HC are lower in case of B20 among all the blends and diesel. And very close to each other at full load CO is lower in case of B20 among all the blends and diesel	The ID of all blends is longer. However, in comparison to standard diesel fuel, B20 has a higher cylinder pressure The HRR is also higher Increment in EGT with increase in load from low to full. B20 shows

(continued)

Table 1 (continued)

Author	Engine used	Performance	Emission	Combustion
		For all loads, the BTE is lower than that of diesel. B20 has good BTE among all blends but still lower than that of diesel	The smoke density increases as the load increases. B20 shows lower smoke density as compared to other blend NOx emanations are lower in case B20 among all blends but greater than that of diesel fuel at full load	lower EGT among all blends at full load but still greater than that of diesel
Hernández et al. [27]	Light-duty engine	–	Most NO_x produced during high part load Higher NO_x with MOBio. Lower NO_x with HVO After LNT MOBio had the worst efficiency since EGT and EGR were similar for fuels So the confirmed effect of fuel type is very important	–
Agrawal et al. [28]	Single-cylinder diesel engine	0 to 20% share of linseed in blend leads to increased BTE. At 30% it is very low and 30 to 50% share leads to an increase in BTE. At 50% share the BTE is maximum	–	Cylinder pressure (CP) was max. for a 50:50 ratio blend From 0 to 30% fraction of linseed oil leads to decrease in CP and 30 to 50% leads to increase HRR of 50% fraction blend is greater than all blends and diesel EGT of 50% fraction blend is higher and other for other blends are quite close to diesel

(continued)

Table 1 (continued)

Author	Engine used	Performance	Emission	Combustion
Tongroon et al. [29]	Single-cylinder diesel engine	Among all blends torque and power are higher for B7E5 but lower with concern of diesel BSFC is similar for B3E5 and B7E5 but higher than diesel	The HC and NO_x levels increased as the ethanol percentage in the blend is increased, surpassing diesel. CO emissions are lower than measuring range	With ethanol fraction increasing, the blends induced higher HRR than that of diesel due to long ID Cylinder pressure is lower than diesel
Karthikeyan et al. [30]	Single-cylinder CI engine	BSFC at full load, highly decreased for B20 and gives the best results among all blends and diesel At full load 20% fraction biofuel blend (B20) gives the best BTE among all blends and diesel	–	With increasing loading conditions and blend proportion, EGT rises. At full load, EGT is higher for B20 among all blends and diesel
Karthikeyan et al. [31]	Four-stroke single-cylinder diesel engine	Enhancement in engine performance and reduction in BSFC	As loads rise from low to full, the HC emissions also raise but decrease in Al_2O_2-based biodiesel and minimum for B20 + 0.2 Al_2O_2 at full load Similar behavior for CO NO_x is higher in all Al_2O_2-based blends and increases with increase in concentration of Al_2O_2 in blend Smoke is reduced for Al_2O_2/B20 as concentration of Al_2O_2 in blend raises, as compared to B20 at full load	–

(continued)

Table 1 (continued)

Author	Engine used	Performance	Emission	Combustion
Aravind et al. [32]	Single-cylinder four-stroke diesel engine	BSFC decreases from low to high load for all blends and B20 shows best results among all blends BTE rises from 0 to 75% load and higher at 75% after that it declines at 100% load for all blends BTE is lower than diesel for all blends but at 75% load B20 shows very close BTE as that of diesel	CO, HC emanations are lower for all blends than that of diesel NO_x is lower for B20 still higher than diesel	At all blends exhaust gas temperature is low as compared to diesel. B30 shows lowest and B10 shows highest EGT among all blends
Hossain et al. [33]	Indirect injection compression ignition engine	BSFC for emulsified blends and neat rapeseed oil are higher than that of diesel for E3 at full load BSFC is very close to diesel BTE rises as load rises. For all loads, neat rapeseed oil gave higher BTE among all fuels At full load, the BTE is higher for E1 blend and lowest for E3 among all blends and diesel BSFC decreases as load increases	Higher the load higher the CO_2 emission. For E3 the results are almost similar but for other emulsified blends CO_2 is higher than that of diesel Similar behavior is observed for CO as of CO_2 As load increases low to full NO_x increases. All emulsified blends give lower NO_x than that of diesel The amount of smoke generated increases as the load increases. Both blends and neat rapeseed oil emit less smoke than diesel under all load conditions	In general, the EGT is lower for all emulsified blends but at full load the EGT were similar as of diesel

(continued)

Table 1 (continued)

Author	Engine used	Performance	Emission	Combustion
Kumar et al. [34]	Single-cylinder four-stroke diesel engine	If the load increases, the BSFC decreases. In both blends, the BSFC is smaller than diesel while utilizing nanoparticles BTE ascends as engine load rises. In both blends, the BTE is higher than diesel and if nanoparticle is being used the outcome was far better	HC emissions are reduced in case of both blends as compared to diesel. And the case of nanoparticle blending gives more reduction in HC If there is an occurrence of CO and NO_x emanation, similar behavior is seen except HC emanations	–
Venkatesan et al. [35]	Twin cylinder CI engine	BSFC for all blends is higher than that of conventional diesel at low load and became lower at higher engine loads. 50:50 ratio of pine oil and soapnut gives the best results As load rises the BTE improves. At full load, the BTE is highest for neat pine oil and higher than diesel for P50SNB50 and P75SNB25 blend P75SNB25 proved to be a potentially good substitute	40% reduction in UHC for P50SNB50 and P75SNB25 blend with relative to standard diesel at peak load Decrease in CO emission by 6% for P75SNB25 at full load than all test fuels At peak load, NOx is higher in case of all blends than neat diesel Smoke emission is less than diesel for all blends except neat soapnut oil up to 75% load and almost the same as that of diesel at full load	No significant change in cylinder pressure for all blends as compared to diesel HRR is higher for neat pine oil than that of conventional diesel and other fuel mix EGT is similar for P50SNB50 and P75SNB25 blend at higher load
Loganathan et al. [36]	Four-stroke, diesel engine	At low to medium load the BSFC is decreased with H_2 addition but increases with addition of EGR at full load the values are very close to one another BTE rises as load rises. H_2 addition increases the BTE and EGR addition decreases the BTE as % of EGR increases	At all load HC decreases with H_2 addition in blend but increases with addition of EGR Similar behavior is observed in CO emission NO_x emission increased with H_2 addition but reduced with EGR addition	Cylinder pressure and HRR are decreased for EGR addition with H_2 enrichment compared to H_2 enrichment without EGR fuel blend at full load condition

(continued)

Table 1 (continued)

Author	Engine used	Performance	Emission	Combustion
Sheriff et al. [40]	Single-cylinder, four-stroke diesel engine	Lemon oil with CeO_2 depicted appreciable BTE with lower value of BSFC among all biofuel blends	The CO and HC emanations for all blends became lower with and without nanoparticles As CNT nanoparticles are added to all biofuels, the NO_x emissions increase Lower smoke emanation with nanoemulsion in all biofuels particularly at low burden	The ignition response is better for nanoparticles with biofuels due to fine atomization conduct EGR is higher for introduction of nanoparticles with biofuels

1. There is a long ignition delay period observed during biodiesel blend usage as compared to diesel fuel. Biofuels typically have a low cetane number, which lowers power output and lengthens the ignition delay time.
2. Cylinder pressure is more in case of biofuel blends as compared to fossil diesel fuel. Cylinder pressure depends on fuel injection timing, advanced timing means high pressure. Huge amounts of fuel ignition during the premixed burning stage prompt an expansion in peak in-chamber pressure.
3. In the case of biodiesel blends, there is a higher HRR. However, HRR basically depends on cetane number, viscosity, the calorific value of the fuel.
4. BSFC is higher in case of biodiesel blends with respect to diesel fuel. It declines with the increase in engine speed and brake power. Lower heating value and higher density incremented BSFC.
5. BTE in some cases is higher and in some cases is lower eventually that depends on the biofuel used. The essential driver of lower BTE is higher viscosity, density than diesel.
6. HC, CO, CO_2, and smoke emanations are lower in case of biodiesel blends as compared to diesel fuel.
7. NO_x emanations in some cases are higher and, in some cases, lower as compared to petroleum diesel fuel. The common cause for the formation of NO_x emanations is mainly the higher in-cylinder temperature. When EGT goes higher, more NO_x produces.
8. When compared to diesel fuel, biodiesel blends produce less particulate matter and soot emissions. In comparison to diesel, the presence of oxygen in biodiesel mixes increased ignition efficiency and reduced soot formation.
9. Smoke opacity and SO_2 emissions are lower for biodiesel blends as compared to diesel fuel.
10. Addition of nanoparticles in the biofuel blends always improves the BTE, reduces the BSFC and emissions but NO_x increases due to better combustion and higher EGT.
11. In biofuel blends, water emulsion increases BTE thus lowering BSFC. On the other hand, a small rise in CO and CO_2 emissions is also observed while all other emissions decrease. A water emulsion is a blend that has a minimal volume of water applied to it. Surfactants may be used to improve the quality.
12. NO_x emissions can be reduced by employing EGR. EGR is a technique in which a small part of exhaust gas is recirculated to intake manifold. A small engine modification is required in this technique. But it badly affects the engine performance and emissions, so optimum fraction of exhaust gas should be used for EGR. Emulsified fuel technique is better than EGR technique.

4 Conclusion

The use of biofuel in diesel engines requires several factors to be considered. The following are some of the findings that can be taken from this present literature review:

1. On utilization of biofuel mix, there is a slight punishment on performance as well as for BTE and BSFC.
2. HC, CO, CO_2 emission reduces up to a certain limit.
3. Particulate matter, soot emissions, and smoke opacity are reduced when biofuel blends are used.
4. Biofuel can be used in the diesel engine in the form of biodiesel blends without any major engine hardware changes.

The problem of lower performance and higher NO_x emission with biofuel can be addressed by nanoparticle addition and water emulsion technique.

References

1. Verma, P., Sharma, M.P.: Performance and emission characteristics of biodiesel fuelled diesel engines. Int. J. Renew. Energy Res. **7**, 245–250 (2015)
2. Dhinesh, B., Lalvani, J.I.J., Parthasarathy, M., Annamalai, K.: An assessment on performance, emission and combustion characteristics of single cylinder diesel engine powered by Cymbopogon flexuosus biofuel. Energy Convers. Manag. **9**, 466–474 (2016)
3. Mofijur, M.G.R.M., Rasul, A.M., Hyde, J., Azad, A.K., Mamat, R., Bhuiya, M.M.K.: Role of biofuel and their binary (diesel–biodiesel) and ternary (ethanol–biodiesel–diesel) blends on internal combustion engines emission reduction. Renew. Sustain. Energy Rev. **14**, 265–278 (2016)
4. Kumar, B.R., Saravanan, S.: Partially premixed low temperature combustion using dimethyl carbonate (DMC) in a DI diesel engine for favorable smoke/NOx emissions. Fuel **11**, 396–406 (2016)
5. Khiari, K., Awad, S., Loubar, K., Tarabet, L., Mahmoud, R., Tazerout, M.: Experimental investigation of pistacia lentiscus biodiesel as a fuel for direct injection diesel engine. Energy Convers. Manag. **9**, 392–399 (2016)
6. Raju, J., Balasubramanian, A., Gajendiran, M., Raj, N.M.: Investigation of performance and emission characteristics using low reactivity fuel and biodiesel blended nanoparticles in diesel engine. Int. J. Eng. Res. **5**(03) (2016)
7. Prasad, V.V.: Performance of 4 stroke diesel engine using coconut oil as biofuel material. Mater. Today Proc. **8**, 5312–5319 (2017)
8. Leite, D., Santos, R.F., Bassegio, D., de Souza, S.N.M., Secco, D., Gurgacz, F., da Silva, T.R. B.: Emissions and performance of a diesel engine affected by soybean, linseed, and crambe biodiesel. Ind. Crops Prod. **6**, 267–272 (2019)
9. Huang, H., Teng, W., Liu, Q., Zhou, C., Wang, Q., Wang, X.: Combustion performance and emission characteristics of a diesel engine under low-temperature combustion of pine oil–diesel blends. Energy Convers. Manag. **10**, 317–326 (2016)
10. Kumar, B.R., Saravanan, S., Rana, D., Anish, V., Nagendran, A.: Effect of a sustainable biofuel–n-octanol–on the combustion, performance and emissions of a DI diesel engine under

naturally aspirated and exhaust gas recirculation (EGR) modes. Energy Convers. Manag. **12**, 275–286 (2016)
11. Annamalai, M., Dhinesh, B., Nanthagopal, K., SivaramaKrishnan, P., Lalvani, J.I.J., Parthasarathy, M., Annamalai, K.: An assessment on performance, combustion and emission behavior of a diesel engine powered by ceria nanoparticle blended emulsified biofuel. Energy Convers. Manag. **9**, 372–380 (2016)
12. Vijayaraj, K., Sathiyagnanam, A.P.: Experimental investigation of a diesel engine with methyl ester of mango seed oil and diesel blends. Alexandria Eng. J. **7**, 215–221 (2016)
13. Han, X., Yang, Z., Wang, M., Tjong, J., Zheng, M.: Clean combustion of n-butanol as a next generation biofuel for diesel engines. Appl. Energy **13**, 347–359 (2017)
14. Piloto-Rodríguez, R., Sánchez-Borroto, Y., Melo-Espinosa, E.A., Verhelst, S.: Assessment of diesel engine performance when fueled with biodiesel from algae and microalgae: an overview. Renew. Sustain. Energy Rev. **10**, 833–842 (2017)
15. Babu, V., Murthy, M.: Butanol and pentanol: The promising biofuels for CI engines—a review. Renew. Sustain. Energy Rev. **21**, 1068–1088 (2017)
16. Ashok, B., Raj, R.T.K., Nanthagopal, K., Krishnan, R., Subbarao, R.: Lemon peel oil—a novel renewable alternative energy source for diesel engine. Energy Convers. Manag. **12**, 110–121 (2017)
17. Srihari, S., Thirumalini, S.: Investigation on reduction of emission in PCCI-DI engine with biofuel blends. Renew. Energy **14**, 1232–1237 (2017)
18. Singh, P., Goel, V., Chauhan, S.R.: Impact of dual biofuel approach on engine oil dilution in CI engines. Fuel **10**, 680–689 (2017)
19. De Poures, M.V., Sathiyagnanam, A.P., Rana, D., Kumar, B.R., Saravanan, S.: 1-Hexanol as a sustainable biofuel in DI diesel engines and its effect on combustion and emissions under the influence of injection timing and exhaust gas recirculation (EGR). Appl. Therm. Eng. **22**, 1505–1513 (2017)
20. Karthickeyan, V., Thiyagarajan, S., Geo, V.E., Ashok, B., Nanthagopal, K., Chyuan, O.H., Vignesh, R.: Simultaneous reduction of NO_x and smoke emissions with low viscous biofuel in low heat rejection engine using selective catalytic reduction technique. Fuel **16**, (2019)
21. Alagu, K., Venu, H., Jayaraman, J., Raju, V.D., Subramani, L., Appavu, P., Dhanasekar, S.: Novel water hyacinth biodiesel as a potential alternative fuel for existing unmodified diesel engine: performance, combustion and emission characteristics. Energy **11**, 295–305 (2019)
22. Hellier, P., Jamil, F., Zaglis-Tyraskis, E., Ala'a, H., Al Haj, L., Ladommatos, N.: Combustion and emissions characteristics of date pit methyl ester in a single cylinder direct injection diesel engine. Fuel **10**, 162–171 (2019)
23. Kamarulzaman, M.K., Hafiz, M., Abdullah, A., Chen, A.F., Awad, O.I.: Combustion, performances and emissions characteristics of black soldier fly larvae oil and diesel blends in compression ignition engine. Renew. Energy **12**, 569–580 (2019)
24. Ogunkunle, O., Ahmed, N.A.: A review of global current scenario of biodiesel adoption and combustion in vehicular diesel engines. Energy Rep. **20**, 1560–1579 (2019)
25. Phasukarratchai, N.: Phase behavior and biofuel properties of waste cooking oil-alcohol-diesel blending in microemulsion form. Fuel **8**, 125–132 (2019)
26. Deivajothi, P., Manieniyan, V., Sivaprakasam, S.: Experimental investigation on DI diesel engine with fatty acid oil from by-product of vegetable oil refinery. Ain Shams Eng. J. **6**, 77–82 (2019)
27. Hernández, J.J., Rodríguez-Fernández, J., Calle-Asensio, A.: When diesel NO_x aftertreatment systems meet advanced biofuels. Res. Eng. **2**, (2019)
28. Agrawal, B.N., Sinha, S., Kuzmin, A.V., Pinchuk, V.A.: Effect of vegetable oil share on combustion characteristics and thermal efficiency of diesel engine fueled with different blends. Thermal Sci. Eng. Progress **8**, (2019)
29. Tongroon, M., Saisirirat, P., Suebwong, A., Aunchaisri, J., Kananont, M., Chollacoop, N.: Combustion and emission characteristics investigation of diesel-ethanol-biodiesel blended fuels in a compression-ignition engine and benefit analysis. Fuel **6**, (2019)

30. Karthikeyan, S., Periyasamy, M., Prathima, A., Sabariswaran, K.: Performance analysis of diesel engine fueled with S. marginatum Macro algae biofuel-diesel blends. Mater. Today Proc. **6** (2020)
31. Karthikeyan, S., Prathima, A., Periyasamy, M., Mahendran, G.: Emission analysis of the diesel engine using Stoechospermum marginatum, brown marine algae with Al_2O_3 nano fluid. Mater. Today Proc. **7** (2020)
32. Aravind, S., Barik, D., Ragupathi, P., Vignesh, G.: Experimental investigation of Lawsonia inermis L. biofuel as fuel in a diesel engine. Mater. Today Proc. **5**
33. Hossain, A.K., Refahtalab, P., Omran, A., Smith, D.I., Davies, P.A.: An experimental study on performance and emission characteristics of an IDI diesel engine operating with neat oil-diesel blend emulsion. Renew. Energy **10**, 1041–1050 (2020)
34. Kumar, M.S., Selvaraj, S.D., Leelakrishnan, E., Vignesh, N.S., Aswini, S.: Influence of curcumin nanoparticle blended biofuel in engine performance, combustion and exhaust emission characteristics. Mater. Today Proc. **5** (2020)
35. Venkatesan, V., Nallusamy, N.: Pine oil-soapnut oil methyl ester blends: a hybrid biofuel approach to completely eliminate the use of diesel in a twin cylinder off-road tractor diesel engine. Fuel **08**, (2020)
36. Loganathan, M., Thanigaivelan, V., Madhavan, V.M., Anbarasu, A., Velmurugan, A.: The synergetic effect between hydrogen addition and EGR on cashew nut shell liquid biofuel-diesel operated engine. Fuel **10**, (2020)
37. Jagtap, S.P., Pawar, A.N., Lahane, S.: Improving the usability of biodiesel blend in low heat rejection diesel engine through combustion, performance and emission analysis. Renew. Energy. **17** (2020)
38. Kotaiah, K., Periyasamy, P., Prabhahar, M.: Performance and emission characteristics of small agricultural diesel engine using Lemongrass oil and its diesel blends. Mater. Today Proc. **05** (2020)
39. Santhosh, K., Kumar, G.N.: Impact of 1-Hexanol/diesel blends on combustion, performance and emission characteristics of CRDI CI mini truck engine under the influence of EGR. Energy Convers. Manag. **16**, (2020)
40. Sheriff, S.A., Kumar, I.K., Mandhatha, P.S., Jambal, S.S., Sellappan, R., Ashok, B., Nanthagopal, K.: Emission reduction in CI engine using biofuel reformulation strategies through nano additives for atmospheric air quality improvement. Renew. Energy **147**, 2295–2308 (2020)
41. Sharma, R., Pal, A., Ansari, N.A.: Utilization of CNG in dual fuel engine: review, WEENTECH. In: Proceedings in Energy **6**(20), 206–220 (2020)
42. Raman, R., Kumar, N.: Experimental studies to evaluate the combustion, performance and emission characteristics of acetylene fuelled CI engine. Int. J. Ambient Energy (2020)
43. Raman, R., Kumar, N.: Experimental investigation to analyze the effect of induction length of diesel-acetylene dual fuel engine. Energy Sources, Part A: Recovery, Utilization and Environmental Effects (2019)
44. Raman, R., Kumar, N.: The utilization of n-butanol/diesel blends in acetylene dual fuel engine. Energy Rep. **5**, 10301040 (2019). ISSN: 2352-4847

Comparative Assessment of the Combustion Property of a Cavity-Based Scramjet Combustor with Strut and Wall Injection Technique

Ravi Kumar Samadhiya and Devendra Singh

1 Introduction

The core technique is supersonic combustion for hypersonic sustainable flights. Usually, the combustion thrust is 1 m long in the generators, the mixture length in milliseconds [1, 2]. The problems of burning chamber arise from high supersonic speeds, reagent mixing, flame anchoring, and stability and combustion success during low combustion duration. The flow field in combustion of the scramjet is very complex [3]. When the flight speed is low, the kinetic energy in the air must not be used for optimum compression. To obtain a better efficiency, additional device in a compact form is required. The turbojet, for example, uses an extra compressor engine. If the flight speed reaches a certain level, after maximum compression, the air flow into the combustion tank is supersonic [4]. With extra compression, the engine performance will decrease (i.e., deceleration). Combustion should also occur in the supersonic flow. The so-called 'supersonic ramjet combustion jet' is an air breathing motor that acts under hypersonic flight conditions, where the term 'supersonic combustion' refers to combustion in the case of supersonic flux [5].

Scramjet is an air respiration vehicle that is most promising and can be used under supersonic and hypersonic conditions [6, 7]. Because of the ram effect, the zero flight speed is no thrust. It is therefore advantageous to run the engine with fixed geometry both in scramjet and in ramjet modes [8]. A numerical analysis was conducted on mixing and combustion improvement [9] for Mach 2. The exceptionally short duration of air residency in supersonic fuels prevents efficient combustion of fuel and air (quick and with limited losses in overall pressure). The effect of chemical simulation on supersonic predictions of hydrogen combustion is being investigated in the prototype commutator. In single-stage chemical studies,

R. K. Samadhiya (✉) · D. Singh
Department of Mechanical Engineering, Sachdeva Institute of Technology, Farah, Mathura, India

L. M. Das et al. (eds.), *Recent Trends in Thermal Engineering*,
Lecture Notes in Mechanical Engineering,
https://doi.org/10.1007/978-981-16-3428-4_15

"multi-stage chemical ventures are expected to provide a higher and broader spread of heat". Furthermore, the ignition distance and induction distance are examples of multitasking chemical engineering that foresee dynamic aspects of the combustion phase.

The collected supersonic combustion data is suitable for powerful, low-status scramjet motors. The methods used to ignite the fuel and to preserve combustion are given consideration. This is especially evident in supersonic combustion experiments, and some examples are seen in tests at far higher inlets than in flight temperatures. The explanation is good: it is impossible to maintain a low temperature supersonic flow of hydrogen or hydrocarbon. A correctly built combustor that can be used to run a less effective combustor merely increases the static temperature to spontaneous ignition makes this possible. The low input/basic temperature is ideal because of the intake and nozzle constraints.

A Lagrangian high speed turbulent combustion model is used, combined with a popular RANS-AMA method, to model non-reactive, non-reactive, turbulent supersonic co-flow jets. C has been tested experimentally for hydrocarbon supersonic combustion. Mach 2.15—The air from the scramjet combuster model has burned kerosene continuously. The fuel is injected into the pylon in the supersonic airflow. In the kerosene process the liquid fuel jet is dispersed and a small amount of hydrogen is added. This additional jet dispersion of supersonic combustible is not needed until the fuel is usually pumped into crossflows. The aerial flow temperature must be lowered under Tmin's combustion ignition stage before the fuel is encoded. The time/residence rate below T reaches its volatility again and the fire dies. The fire is not exposed anymore. Figure 1 diagram representing schematic view of scramjet engine.

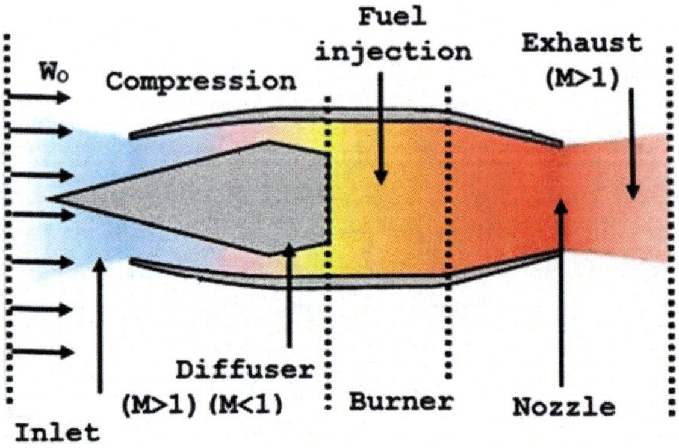

Fig. 1 Diagram representing schematic view of scramjet engine [12]

2 Approach to the Problem

The supersonic combustion system based on a cavity is shown in Fig. 2. In the experiment results of Micka et al. [12–14], all dimensions and boundary conditions have been adopted. The supersonic combustible comprised of an insulator and combustion device with a cavity flame holder of 52.6 mm and 13.4 mm in length and depth. The height and length of an insulator at the cavity edge are 28.8 and 410 mm at a different angle of 45.

The strut is positioned on the center line of the combustor, at a distance of x 360 mm and y 13.6 mm. For both walls and strut injection systems, fuel was injected at x 363 mm. Takes the strut dimensions as shown in Fig. 3.

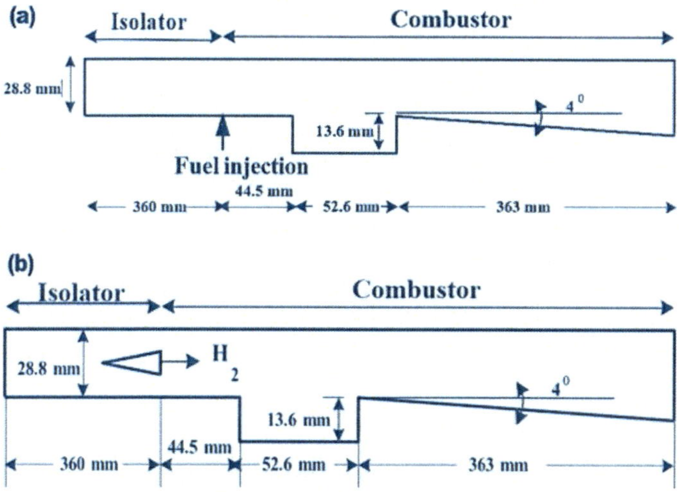

Fig. 2 Schematic diagram: **a** scramjet combustion cavity-based [17–19], **b** scramjet combustor with a strut and cavity

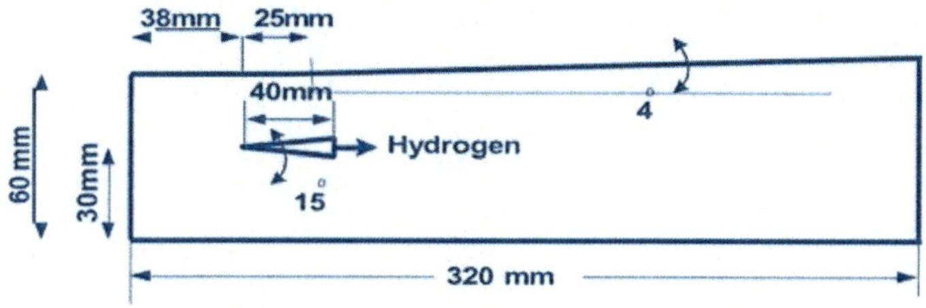

Fig. 3 DLR-based combustor schematic diagram

2.1 Boundary Conditions

The boundary conditions of Neumann and Dirichlet for the factors in outlet and inlet constraints are used in this research. No-slip limits on fixed walls are nevertheless used. Table 1 shows the limits of free flow air and H2 jet.

3 Analytical and Experimental Formulations

The section investigates the combustion process in a cavity-based scramjet combustion process for stretches and walls [15]. The fluctuation is regarded as two-dimensional and stable. The following regulatory numerically solved transportation equations with fluent software from ANSYS 16.0. For volumetrically responses, the implicit density-based solver and finite dissipation model are used. Mass, momentum, and energy conservation equations were expressed in the numerical solution of the dimensional forms of the governing équations.

3.1 Governing Equations

In addition to inquiries in a wide variety of flying conditions and geometries, it is also possible to reliably determine the advantages of the whole Navier-stock equation as regards the direction of shock. Below are the three-dimensional shapes of the equations of Navier-Stokes. Note that only in the coordination direction, the two-dimensional forms are generalized of the three-dimensional equations, due to lack of factor variables. The three-dimensional Navier-Stokes equations have been derived as a neglect of the nature of body forces and volumetric heating [16].

$$\frac{\partial p}{\partial t} + \frac{(pu)}{dx} + \frac{(pv)}{\partial y} + \frac{(pw)}{\partial z} = 0 \tag{1}$$

Table 1 Parametric considerations for this research

S.No.	Parameter	Air	Hydrogen
1	Ma	3.0	1.0
2	$T0$ (K)	1310	290
3	$P0$ (kPa)	610	150
4	XO_2	0.21	0
5	XN_2	0.58	0
6	XH_2O	0.21	0
7	XH_2	0	1

$$\frac{\partial(pu)}{\partial t}+\frac{\partial(puu)}{\partial x}+\frac{\partial(vu)}{\partial y}+\frac{\partial(wu)}{\partial z}=\frac{\partial\sigma_{xx}}{\partial x}+\frac{\partial\sigma_{yx}}{\partial x}+\frac{\partial\sigma_{zx}}{\partial x} \tag{2}$$

$$\frac{\partial(pv)}{\partial t}+\frac{\partial(puv)}{\partial x}+\frac{\partial(vv)}{\partial y}+\frac{\partial(wv)}{\partial z}=\frac{\partial\sigma_{xy}}{\partial x}+\frac{\partial\sigma_{yy}}{\partial x}+\frac{\partial\sigma_{zy}}{\partial x} \tag{3}$$

$$\frac{\partial(pw)}{\partial t}+\frac{\partial(puw)}{\partial x}+\frac{\partial(vw)}{\partial y}+\frac{\partial(ww)}{\partial z}=\frac{\partial\sigma_{xz}}{\partial x}+\frac{\partial\sigma_{yz}}{\partial x}+\frac{\partial\sigma_{zz}}{\partial x} \tag{4}$$

$$\begin{aligned}\frac{\partial(pE)}{\partial t}&+\frac{d(puE)}{dx}+\frac{d(pvE)}{dy}+\frac{d(pwE)}{dz}\\&=\frac{\partial\left(u\sigma_{xx}+v\tau_{xy}+w\tau_{xz}\right)}{\partial x}+\frac{\partial\left(u\sigma_{yx}+v\tau_{yy}+w\tau_{yz}\right)}{\partial x}\\&\quad+\frac{\partial\left(u\sigma_{zx}+v\tau_{zy}+w\tau_{zz}\right)}{\partial x}+\frac{\partial\left(k\frac{\partial T}{\partial x}\right)}{\partial x}+\frac{\partial\left(k\frac{\partial T}{\partial y}\right)}{\partial x}+\frac{\partial\left(k\frac{\partial T}{\partial z}\right)}{\partial x}\end{aligned} \tag{5}$$

If the Newtonian liquid is assumed, the normal strain $\mu xx'' \pm y'$ and $\acute{S}zz$ can be used as a pressure p combination with the normal $\mu xx'\mu \pm Yy'$ and μjzz stress components, while the rest are viscous stress components which are tangential elements $\tau xy = \tau yx$, $\tau xz = \tau zz$= and $\tau yz'$, = $\tau zy'$ Specific energy E is resolved for the energy efficiency of supersonic flows instead of the usual one. The true energy can be expressed as E mentioned below in the formulae

$$e+(u2+v2+w2) \tag{6}$$

Because of the high velocities that currents can achieve when Ma > 1, it is clear that the principle of kinetic energy plays a significant role in energy conservation. Equations (1)–(6) denote the form of regulatory equations that are used in the case of compressible flows.

The scramjet fundamental principle is that of the Jet engine, where the major organizational structures are exhaust, compression, combustion, and exhaust. Yet a ramjet has no moving components, no moving pieces and no moving parts. The air stream is distorted by the compression effect of the ram. The subsonic intake air velocity decelerates the subsonic velocity [17, 18]. Tanks are needed subsonic to the combustion phase. A mechanical shock or a short throat is available for the exhaust materials. However, if the flow is a scramjet, it never decreases subsonic speeds. This involves burning at supersonic speeds. The entire propulsion system consists of five in the scramjet. Main components of engine and two components of trucks, the engine consists of internal inlet, insulator, combustion engine, and internal inlet, subsystem of fuel supply and compressor, forebody and afterbody two vehicle components, and if the scramjet is of technically simple appearance, several issues persist waiting for acceptable solutions involving conceptual shift simulation of wave encounters, turbulent mixing, two-stage flow, flow shielding, burning of strongly reactive species, real gas aerodynamics, and components [19–22].

4 Result and Discussions

The numerical simulation with 691,261 elements, 300,125 elements, and 564,221 nodes was performed. For each combustor geometry, three different mesh sizes have been considered.

The static pressure distribution is described in Table 2 along the downstream combustion length. The *x*- and *y*-direction mesh distribution along with the strut location are shown in Fig. 4. It is found that, even with the growth of the number of elements from 300,125 to 498,225, the static pressure distribution is nearly identical. Therefore, for the strut injection system 271,542 elements are required.

Figure 5 shows a viewpoint image of the responding flow and a qualitative comparison of H2 injections (1.50 g/s) and combustion experimental and numerical shadow graphic photos. The expansion fans at the upper and lower ends of the edge are effectively vanishing with H2 injection and combustion. The re-compression shocks are smaller than inert H2 injection situations. The recirculation zone is longer and wider with combustion and is used as flame retardant to recover H2 diffusion fires, LES peak velocity reverse (recirculation area), and an increase in the basis pressure in relation to the experiment-supportive injection case of H2[30]. A transition zone characterized the broad and consistent flow structure, convective mix, air induction, and exothermic induction, and a turbulent combustion zone can be divided into roughly three regions with large-scale consistent structures and combustion: the flow zone where mixing and progression of burning are determined by the turbulence. The wide structures arise from the shear layers which roll up and gradually twist downstream due to a downturn in the vortex, which happens mostly due to the dilation induced by the effluvium and the exothermic effect of the combustion. The shocks interact with exothermic architecture increases the shear thickness and affects subsequent shocks and reflection of the chamber pressure. The broad structures trigger dynamic exchange by the mixture of cool, high energy air with H2 heat or substance wake flow, enabling the transition from the sound wake flow to supersonic and free flow. We see just the right agreement on the agreement between the foretold and experimental shadow graphics.

Table 2 Test static pressure on the bottom wall of the combustion system at various sizes of grid with wall and strut injection system

Wall injection			Strut injection		
No. of elements	Static pressure (kPa)	% error	No. of elements	Static pressure (kPa)	% error
61,921	150	4.2	56,258	220	4.6
300,125	160	0.0	271,542	230	0.0
564,221	160	0.0	498,225	230	0.0

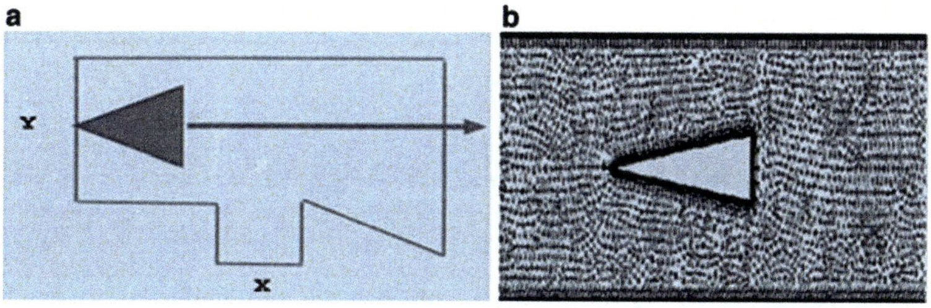

Fig. 4 **a** Details of non-uniform grid data used, **b** prolonged view at the strut

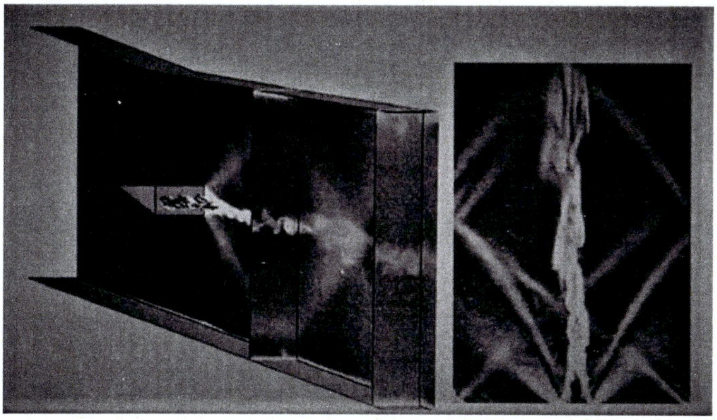

Fig. 5 Reaction case: **a** view of axial momentum contours of the back plane and of the two planes, temperature contours, flame surface isolation and H2 mass fraction iso-surface, and **b** the numerical shadow map (essentially r2q)

4.1 Modeling of Combustion

Finite rate eddy dissipation model was used in this paper in comparison to the eddy dissipation template for more precise results in the calculation of the reaction flow field. To minimize computational time, the CFL number was set to 0.6 and a single-step chemical reaction was selected. The single-step reaction can also be written in the following form. Figure 6 shows that the fuel efficiency is the ratio of the hydrogen mass flow rate in one section to the injector mass flow rate

$$2H_2 + O_2 \rightarrow 2H_2O$$

The fuel efficiency is the ratio of the hydrogen mass flow rate in one section to the injector mass flow rate and can be stated as follow.

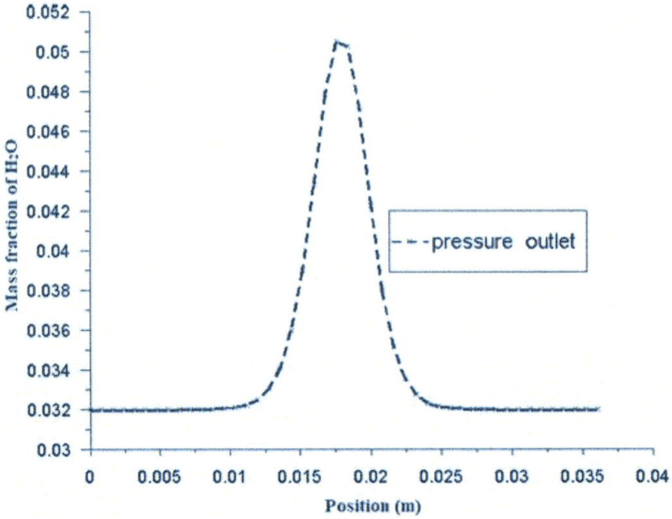

Fig. 6 Fuel efficiency is the ratio of the hydrogen mass flow rate in one section to the injector mass flow rate

4.2 Grid Independence Test (GIT)

The numerical simulation with 691,261 elements, 300,125 elements, and 564,221 nodes was performed. For each combustor geometry, three different mesh sizes have been considered (Fig. 7).

The illustration above represents initial shadow graph of unreacted event, a get about defined as a broad and coherent flow system, a convective mixture, air induction, and exothermic induction as well as a turbulent combustion zone can be described into three regions with a big, uniform structure, and combustion: liquid, flow, and non-turbulent combustion zone: The shocks interfere with an exothermic architecture, increase shear thickness, and influence the subsequent shocks and reflexes in the chamber. Broad structure dynamic exchange of cold, potent air, and H2 heat or wake-flow materials to convey sound through supersonic and free-flowing materials dynamic exchange. We will see that the graphics predicted and experimented have the right agreement. Measured images here as well as in the wake Volumetric development Right off the wedge due to reactions in the transitional zone. But it cannot be clarified entirely why: I have poor mixing of sub-grid models; (ii) typical use of a two-step, too quick solution; (iii) great gas thermal model, not thermally perfect gas model; and (iv) misusing the shadow images of r2q. The reasons for this cannot be fully explained. The Oh-LIF dispersion observations, on the other hand, were checked with proven LES results if the volumetric expansion effect was much less pronounced than the experimental shadow. The researchers do not know if the results of the sample are very unclear.

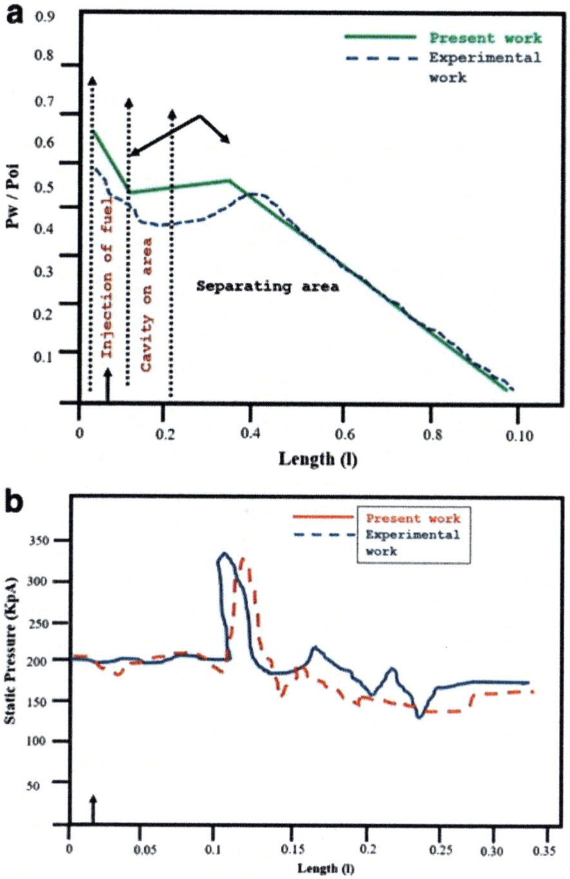

Fig. 7 **a** For a cavity-based combustor, static mean wall pressure distributions, **b** a strut-based DLR combustor's static pressure distribution

4.3 CFD Analysis

The shock wave chart and its temperature contour are shown in Fig. 8. The fluids acceleration behaviors behind the strut were seen when the strut was applied. Diamond obligate shock waves start at the front end of the strut that passes through the burning chamber.

The resulting diamond-shaped flow injector is shown in Fig. 8. The lower and top part of the combustion walls, which hits the wake where a large amount of fuel jet mixes with supersonic air, are reflected off on the border wall of the combustion chamber shock waves. For diamond-shaped injectors, the optimum temperature is 3186.35 K, and this is a very high combustion temperature.

The injector port controls the wall injection system by the bow shock wave, and the bow shock waves type V can be detected in the combuster section. In case the fluid flow disruption was apparent in the case of a strut injection system to help create more edges and vortices, this will probably be mixed more often because of

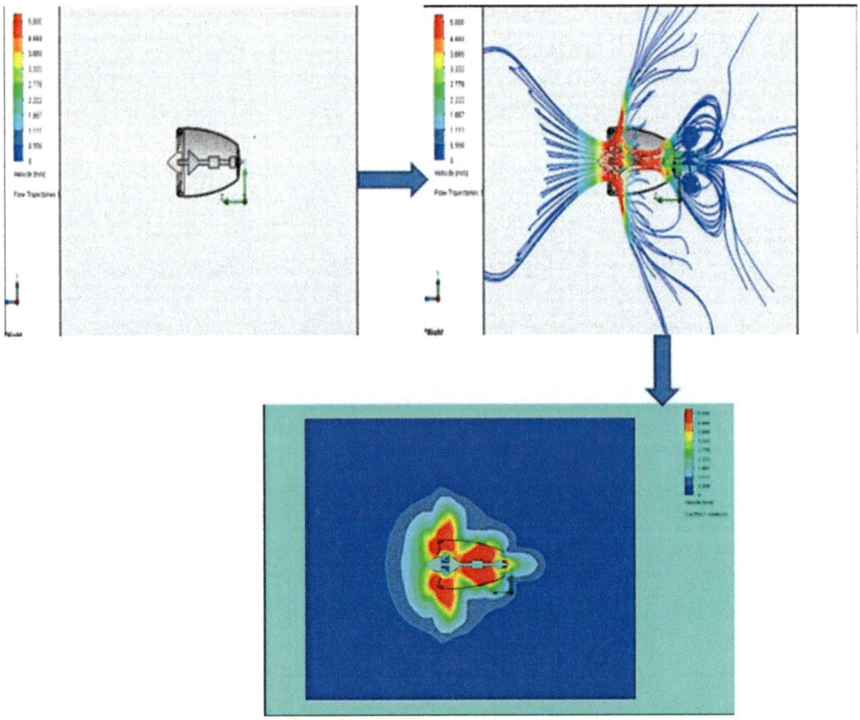

Fig. 8 Starts in a near-wall region and causes wall damage with high temperatures, pressures, and shock waves

the edges and vortex. Figure 8 starts in a near-wall region and causes wall damage with high temperatures, pressures, and shock waves. The combustion system also starts with a combustion system that allows the fuel efficiency in relation to the wall injector to be mixed and refined.

5 Conclusion

- Primary focus is the strength of the local thermal escape, specifying the methods of inflammation and gas stabilization, as well as the geometry of the pipeline and the injection techniques and the accuracy of the fuel mixture to the oxidizer.
- Some regions to be further concentrated, including the total ambient energy losses for combust, are planned to allow different supreme currents and dynamic viscosity to be controlled. Using a CFD method to use a range of fuel and combustion chamber results, pressure and optimal activity isolation. These devices are identical to systems that do not have H2 injection.

- This gives the region a typical shock wave pattern downstream from the bottom of the bow, next to the gently bent expansion fan. This local improvements include thickening of the edge layer, increased fluctuations.
- Discrete boundary layers are on the top of the coin and a shear sheet on the foundation. This shear layer is of course volatile and can therefore be broken into Kelvin-Helmholtz structures (KH). The wake loaded with the H2 is downstream rather than the lowest shock when it pushes a single side tube from one direction.
- Not only are geometry influenced by the combustion phenomenon and wave structure, but the flight Mach and combustion trajectory also affect. Since the time of mixing fuel/air is about a millisecond, it is critical to create better injection systems that improve fuel/air mixing. The regional combustion regions with a high heat capacity in combustible systems are formed due to the shockwave, thereby improving the combustion phenomenon.
- The main focus is on the low intensity of heat, the control of flame starter and equilibrium methods, fuel injection processes, air supplied quantity, fuel-to-air mixing, and gas-dynamic stream framework, along with duct geometry. A number of variables depend on the required combustion residence time. The hypothetical quantity of the combustion chamber is based on the flow rate of the propulsion systems, the minimum combustion product density, and the durability of the combustion process.

References

1. Tretyakov, P.K.: The Problems of Combustion at Supersonic Flow, West-East High Speed Flow Field Conference, vol. 19, 22, November 2007
2. Aso, Shigeru, Hakim, Arifnur, Miyamoto, Shingo: Kei Inoue And Yasuhiro Tani—Fundamental study of supersonic combustion in pure air flow with use of shock tunnel. Acta Astronaut. **57**, 384–389 (2005)
3. Cain, T., Walton, C.: Review of experiments on ignition and flame holding in supersonic flow. Published by The America Institute of Aeronautics and Astronautics, Rto-Tr-Avt-007-V2
4. Mo Bus, H., Gerlinger, P., Bru, D.: Ggemann-scalar and joint scalar-velocity-frequency Monte Carlo Pdf simulation of supersonic combustion. Combust. Flame **132**, 3–24 (2003). PeterGerlinger
5. Stoll, P. 1, Kindler, M., Fernando Schneider C., Aigner, M.: Numerical investigation of mixing and combustion enhancement in supersonic combustors by strut induced streamwise vorticity. Aerosp. Sci. Technol. **12**, 159–168 (2008)
6. Fry, R.S.: A century of ramjet propulsion technology evolution. J. Propul. Power **20**(1), 27–58 (2004)
7. Curran, E.T.: Scramjet engines: the first forty years. J. Propul. Power **17**(6), 1138–1148 (2001)
8. Torrez, S.M., Driscoll, J.F., Ihme, M. and Fotia, M.L.: Reduced-order modeling of turbulent reacting flows with application to ramjets and scramjets. J. Propul. P. **27**(2), 371–382 (2011)
9. Manna, P., Chakraborty, D.: Numerical simulation of transverse H2 combustion in supersonic airstream in a constant area Duct **86** (2005) ARTFC

10. Huang, W., Wang, Z.G., Yan, L., Liu, W.D.: Numerical validation and parametric investigation on the cold flow field of a typical cavity-based scramjet combustor. Acta Astronaut. **80**, 132–140 (2012)
11. Huang, W., Wang, Z.G., Pourkashanian, M., Ma, L., Ingham, D.B., Luo, S.B., Lei, J., Liu, J.: Numerical investigation on the shock wave transition in a three-dimensional scramjet isolator. Acta Astronaut. **68**(11–12), 1669–1675 (2011)
12. Micka, D.J.: Combustion stabilization, structure, and spreading in a laboratory dual-mode scramjet combustor. Ph.D. Thesis, University of Michigan (2010)
13. Micka, D.J., Driscoll, J.F.: Combustion characteristics of a dual-mode scramjet combustor with cavity flame holder. Proc. Combust. Inst. **32**(2), 2397–2404 (2009)
14. Micka, D., Driscoll, J.: Dual-mode combustion of a jet in cross-flow with cavity flame holder. 46th AIAA Aerospace Sciences Meeting and Exhibit, pp. 7–10 (2008)
15. Aleksandrov, V.Y., Prokhorov, A.N., Semenov, V.L.: Hypersonic technology development concerning high speed air-breathing engines. In: 10th International Congress of Fluid Dynamics, AinSoukhna, Red Sea, Egypt, pp. 1–15, 16–19 (2010)
16. Deepu, M., Gokhale, S.S., Jayaraj, S.: Recent advances in experimental and numerical analysis of scramjet combustor flow fields. J. Inst. Eng. 88, 13–23 (2007)
17. Gerlinger, P., Bruggemann, D.: Numerical investigation of hydrogen strut injections into supersonic airflows. J. Propul. Power **16**(1), 22–28 (2000)
18. Pandey, K.M., Roga, S.: CFD analysis of scramjet combustor with non-premixed turbulence model using ramp injector. Scientific.Net Appl. Mech. Mater. **555**, 18–25 (2014)
19. Pandey, K.M., Roga, S., Choubey, G.: Computational analysis of hypersonic combustion chamber using strut injector at flight mach 7. Combust. Sci. Technol. **187**, 1392–1407 (2015)
20. Roga, S., Pandey, K.M.: Computational analysis of hydrogen-fueled scramjet combustor using cavities in tandem flame holder. Scientific.Net Appl. Mech. Mater. Switz. **772**, 130–135 (2015)
21. Rabadan, E., Weigand, B.: Numerical investigation of inlet-combustor interactions for a scramjet hydrogen-fueled engine at a flight Mach number of 8. In: 18th AIAA/3AF International Space Planes and Hypersonic Systems and Technologies Conference, Paper ID: AIAA-2012-5926, Tours, France (2012)
22. Pandey, K.M., Roga, S.: CFD analysis of hypersonic combustion of H2-fueled Scramjet combustor with cavity based fuel injector at flight mach 6. Scientific.Net, Appl. Mech. Mater. **656**, 53–63 (2014)

Study of Eco-friendly Combined System for Refrigeration and Power Generation

Ranjeet Kumar Singh, Tarun Kumar Gupta, Vinod Kumar Yadav, Prabhat Ranjan, Pankaj Kumar Chauhan, and Akash Deep

1 Introduction

Heat liberated during the chemical reaction or fuel combustion process is known as waste heat. This liberated heat is discarded into the atmosphere despite the fact that it could be used again for beneficial and cost effective purposes. Here important matter of fact is not the amount of heat liberated rather the quality of heat in form of its "value." Waste heat is liberated throughout a plant during the manufacturing process of different industries. Hot air of heating systems like gasses (at high temperature) coming out from burners during the process heating; waste heat of flue and exhaust gasses; gasses (at lower temperature) coming out from furnaces during heat treatment process, heaters, and dryers; and heat of cooling liquids, heat exchangers, and gasses are the prime sources of waste heat. However, waste heat of exhaust gases is acknowledged readily, but waste heat in liquids like hot wash water, cooling tower and blow down water, and heat in solids like finished hot products after completion of manufacturing process or after completion of reactions, heated by-products during combustion of solid materials or after processing can also be source of waste heat.

In spite of it, other sources of waste heat as steam-leaks, hot surfaces, and boiler blow down water are not apparent.

Waste heat gases outlet temperature and economics involved are the two main factors to decide the approach of how to recover this heat. By recovering this waste heat, a significant amount of fuel can be saved on primarily basis. Although, waste heat energy cannot be entirely recovered, yet considerable amount of heat could be recovered by adopting different available methods, and losses could be minimized.

R. K. Singh (✉) · T. K. Gupta · V. K. Yadav · P. Ranjan · P. K. Chauhan · A. Deep
Department of Mechanical Engineering, G.L. Bajaj Institute of Technology and Management, Greater Noida, India
e-mail: ranjeetsing@gmail.com

L. M. Das et al. (eds.), *Recent Trends in Thermal Engineering*, Lecture Notes in Mechanical Engineering,
https://doi.org/10.1007/978-981-16-3428-4_16

Based on the process, waste heat discharge can be at different temperature levels in chilled cooling towers to a high temperature levels in industrial furnaces. Generally, higher the exhaust gases temperature and higher the waste heat quality ensure the effectiveness of recovered heat in terms of cost. During the waste heat recovery analysis, recovered heat utilization is absolutely necessary.

Various waste heat recovery (WHR) devices like recuperator, regenerator, heat wheel, heat pipe, economizer, heat pump, etc., are known and available for uses. Durability, silence, scalability, and small size are some desirable characteristics of thermoelectric generator (TEG) waste heat recovery system, when it is compared to other similar technologies. TEG waste heat recovery system neither has chemical reaction nor the moving parts; hence, requirement of maintenance is low due to no wear and corrosion. However, TEG waste heat recovery system has low efficiency relative to a Rankine cycle waste heat recovery system; but due to the fact of no cost associated with waste heat, efficiency is not the only and most important factor. All the mechanical parts of thermoelectric refrigeration system are eliminated and replaced by thermoelectric module. Nakhate et al. [1] conducted various experimental work on eco-friendly refrigerator using Peltier device. During his experimental work and investigation, the 12 V power supply was provided through a switch mode power supply(SMPS) Peltier plate and heat sink fan module, and they did not utilize sun-oriented energy for an inventory unit of the task consequently an appropriate temperature range for transitory food stockpiling is 3–5 °C (37–41 °F), with the utilization of microcontroller planned a criticism unit for a unit where the holder is cooled to required temperature by then inventory naturally turn off. Nandini [2] have analyzed Peltier-based cabinet cooling framework utilizing a heat pipe and fluid-based warmth sink as well as heat pipe. The hot side of both the Peltier device is appended to a similar fluid-based warmth sink, and the virus side is joined to the warmth pipe-based warmth sink. The framework has ten blades on the top and base side of the cover plate. The cover plate was planned and created utilizing aluminum material. Dhumal et al. [3] have work on thermoelectric refrigeration system based on solar energy. A mathematical model has been developed to verify experimental result. The capacity of refrigerator is 0.5 l. Observation reveals that 24 °C is decreasing in cold space after 7 min of running. Pushkarny et al. [4] have studied and analyzed solar cell that develops 17 V and 1.16 A current at 20 W power supply. The cooling space of refrigerator is 7.8 L which has required 60 W power supply to maintains their temperature. Gandhewar et al. [5] used thermoelectric module and fabricated a solar operated heating and cooling system using the same module. It avoids any unnecessary electrical hazards and proves to be environment friendly.

This paper is based on the principal of Seebeck effect. In this paper, the waste heat is converted from a Peltier refrigerator. The waste heat is used to generate the electricity. TEG and TEC are used as a main component and heat sink, cooling fan, and insulation materials also. This work is totally dependent on the difference of temperature. As the temperature difference increases, the electricity generation will increase. Normally in refrigerator and air conditioner, there is an excess amount of heat get rejected which has no use. This heat or waste heat can be used for generation of electricity after sufficient temperature rise or the significant difference between the atmospheric temperature and rejected heat temp. Here with the help of TEG, the temperature difference converts into useful electricity.

2 Materials for Eco-Friendly Refrigerator

The following is the list of material required for manufacturing the refrigerator:

A **Solar panel**
Solar panel can generate electricity by two means. One is by using thermal method and other is by using photovoltaic effect. In this work, the photovoltaic effect has been used for direct conversion of solar energy into electricity, i.e., the conversion of light or other electromagnetic radiation into electricity without any other conversion. Heat can be converted directly into electrical energy by solar cell, more generally a photovoltaic cell. The solar panel use in this fabrication provides up to 10 W of clean, free, and renewable power.

B **Charge controller**
To prevent over charge from wind generators and solar panels, the charge controller is used. It is efficient and simple in design. It provides outputs drive current of intensity 1.0 A.

C **Battery**
To convert chemical energy into high grade energy, i.e., electrical energy, batteries are used. In the present investigation, a 12 V lead acid battery is used.

D **Thermoelectric module**
Selected thermoelectric module has two substrates of ceramic that works as a base and non-electrical conducting material for N-type and P-type dice that are jointed electrically and thermally in series and parallel, respectively, between the ceramics. Further, ceramics also work as insulation between the modules and heat sink. A thermoelectric generator (TEG) and thermoelectric chip (TEC) are also used in this work.

E **Heat sink**
The heat sink shown in Fig. 1, made by aluminum, is in contact with the hot side of a thermoelectric module. Dimensions of heat sink are shown in Table 1. When the positive and negative module leads are connected to the respective positive and negative terminals of a direct current (D.C) power source, heat will be rejected by the modules hot side, and the warmth sink expedites the removal of warmth. Heat sink typically intermediates stages within the heat removal process, whereby heat flows into a conductor and then is transferred to an external medium. Common heat sinks include free convection, forced convection, and fluid cooled. Please note that the first paragraph of a section or subsection is not indented. The first paragraphs that follow a table, figure, equation, etc., does not have an indent, either.

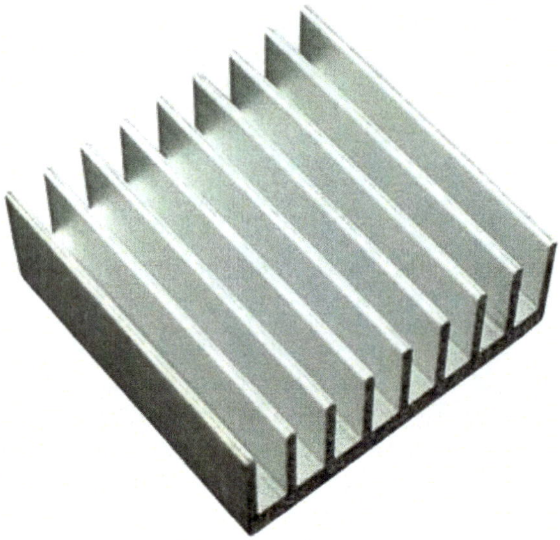

Fig. 1 Heat Sink

Table 1 Specifications of heat sink

Heat sink dimensions	Dimensions in (mm)
Length	138
Breadth	100
Height	18

F **Step up power booster module**

To supply stable power supply system, a power unit (Power booster module) is used. Specification of this unit is as follows.

Weight---13 g Convert---DC 0.9 V to DC 5 V

G **Exhaust fan**

Two fans are used, first one heating fan of diameter 100 mm, and other one is cooling fan of diameter 55 mm

H **Wooden cooling cabinet**

The primary calculations were considered for the device as a rectangular shape because of having a few focal points like its simplicity to construct and insulate. An entryway is appended to one of the sides. At last, any protection, thermoelectric modules, or warmth sinks are effortlessly secured to the sides. CAD model of fabricated cabinet as shown in Figs. 2 and 3 shows prototype of cabinet. A wooden cabinet was used for this purpose having specifications shown in Table 2.

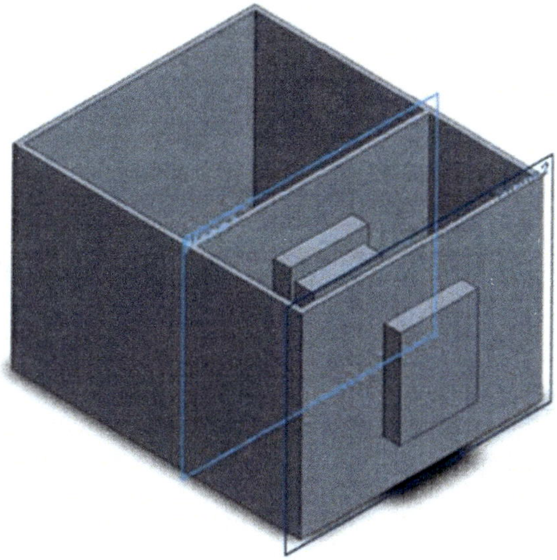

Fig. 2 CAD model of cabinet

Fig. 3 Prototype of cabinet

Thickness of (polystyrene)---1.2 cm

Thickness of plywood ---5 mm

Cooling space $= \mathbf{17.8 * 20.2 * 23.7cm^3 = 8.52l}$.

Table 2 Specifications of wooden cabinet

Outside dimensions	Inside dimensions
Length- 18.8cm	Length- 17.8cm
Breadth- 21.2cm	Breadth- 20.2cm
Depth- 24.7cm	Depth- 23.7cm

3 Result and Discussion

The refrigeration unit can cool itself in seven hours, and the temperature came down in two hours is up to 14 °C. The cooling unit is satisfactorily productive to stay up the temperature through feedback network framework. Temperature sensors help in observing the temperature inside the cooling unit. Planned and built up a thermoelectric cooling framework with a refrigeration space of 8.52-l capacity encased within the box of medium thickness fiber sheet and guarded by polystyrene sheet. Three thermoelectric modules are having been utilized to decrease the temperature of the cold unit. Each thermoelectric module is in touch with particular conductor fans subsequently helping the recent side of the Peltier plate to scatter a greater amount of warmth within the atmosphere.

In 30 min of operation, the significant temperature difference between hot side and atmospheric temperature was obtained as shown in Table 3. Because of this temperature difference across the surface of TEG, it causes the seebeck voltage across the terminals of TEG. Voltage generated by the power booster is up to 5 V, and by this, voltage is sufficient to charge the cell phone.

4 Conclusion

Present work concludes that where ever the heat is being directly released in atmosphere can be converted into useful electricity. It is very useful in reducing the adverse effect of global warming and thermal pollution. Further, enough voltage difference has been generated through 40 °C temperature difference, which is capable to charge a mobile phone. This refrigerator is portable and occupies very less space and can be operated at all places where the heat is wasted. The TEG and TEC can be fabricating according to environment needs and working condition.

Table 3 Experimental observations

S.No.	T_{atm} (°C)	T_{hot} (°C)	$V_{obtained}$ (mV)
1	32.3	60.8	453
2	31.5	65.2	486
3	34.2	75.6	565
4	34.7	88	641

References

1. Nakhate, P., Pawaskar, N., Vatamwar, P., Kalambe, S.: Ecofriendly refrigerator using peltier device. Int. J. Res. Sci. Eng. Special Issue 7-ICEMTE March 2017. e-ISSN: 2394–8299, p-ISSN: 2394-8280
2. Nandini, K.K.M.: Peltier Based Cabinet Cooling System Using Heat Pipe and Liquid Based Heat Sink
3. Dhumal, G.S., Deshmukh, P.A., Kulkarni, M.L.: Experimental investigation of thermoelectric refrigeration system running on solar energy and development of mathematical model. Int. Eng. Res. J. (IERJ) **1**(5), 232–238 (2015). ISSN 2395-1621
4. Pushkarny, B.H., Patel, D., Parulkar, A., Rai, H., Khan, N.: Solar refrigeration using Peltier effect. Int. J. Res. Aeronaut. Mech. Eng. **4**(2), 67–74 (2016). ISSN (Online): 2321-3051
5. Prof. Gandhewar, V.R., Miss. Bhadake, P.G., Mr. Mangtani, M.P.: Fabrication of solar operated heating and cooling system using thermo-electric module. Int. J. Eng. Trends Technol. (IJETT) **4**(4) (2013)

Thermodynamic Investigation of Effect of Variation in Thermo-Physical Properties of Gas Turbine Working Fluid on Cycle Performance

Wasim Akram, Sanjay Sanjay, M. A. Hassan, and Souvik Singh Rathore

Nomenclature

a	Coolant mass flow rate/gas flow rate
$a0, a1, a2, a3, a4, a5$	Constant
c_p	Specific heat at constant pressure (kJ/kg K)
CH_4	Methane
C_2H_6	Ethane
CO_2	Carbon dioxide
Fs_a	Correction factor to account actual blade surface
ΔG_r	Gibbs free energy function (43,890 kJ/kg)
ΔH_r	Lower heating value (kJ/kg)
h	Specific enthalpy (kJ/kg)
ln	Natural logarithm
$\dot{m}$	Mass flow rate (kg/s)
O_2	Oxygen
p	Pressure (bar)
Q	Heat input (kJ/kg)
r_p	Pressure ratio
R	Gas constant (kJ/kg K)
R_c	Cooling factor
s	Specific entropy (kJ/kg K)
S_t	Average Stanton number

W. Akram (✉) · S. Sanjay · M. A. Hassan · S. S. Rathore
Mechanical Engineering Department, National Institute of Technology Jamshedpur, Jamshedpur, India
e-mail: sanjay.me@nitjsr.ac.in

M. A. Hassan
e-mail: hassan.me@nitjsr.ac.in

L. M. Das et al. (eds.), *Recent Trends in Thermal Engineering*,
Lecture Notes in Mechanical Engineering,
https://doi.org/10.1007/978-981-16-3428-4_17

sfc	Specific fuel consumption (kg/kWh)
T_b	Maximum allowable blade surface temperature (K)
T	Temperature (K)
TIT	Turbine inlet temperature (K)
w	Specific work (kJ/kg)
X	Mole fraction

Subscripts

a	Air
Ar	Argon
b	Blade
c	Compressor/coolant
comb	Combustion/combustor
CO_2	Carbon dioxide
D	Destruction
e	Exit
f	Fuel
g	Gas
gt	Gas turbine
H_2O	Water
i	Inlet
m	Mechanical
net	Difference
N_2	Nitrogen
O_2	Oxygen
Pt	Polytropic
Pl	Plant
rat	Rational
t	Turbine

Greek Symbols

ϕ	Thermodynamic property function
ω	Availability per unit mass of gas (kJ/kg)
Ω	Exergy (kJ/kg)
α	Gas flow discharge angle (degree)
ε	Effectiveness (%)
η	Efficiency (%)

Acronym

AFC	Air film cooling
FG	Flue gas

GT Gas turbine
NG Natural gas
DLN Dry low Nox

1 Introduction

Electrical power and energy have been the essence of increase in the quality of life of mankind. With the maturity of fracking in extraction of natural gas, the availability of natural gas has increased significantly. Natural gas is known to be a cleaner fuel as compared to coal, which is primarily used in the production of electrical power utilizing Rankine cycle in India. The increased availability of natural gas has also led to a general decline in its price. Gas turbines are known to deliver highest efficiency while running on natural gas. Data from IEA [1] reveal that power sector presently has an expected market of 290 GW, which is projected to increase to 1100 GW by 2040. With this expected growth in power sector, it is clear that gas turbine is bound to play a major role in providing electrical power to humanity till sustainable energy conversion systems have evolved significantly in terms of maturity of technology, leading to high electrical efficiency and cost per unit of electrical power generator.

Constituents of natural gas includes various combustible and non-combustible gases, i.e., methane, ethane being the combustible components and nitrogen, carbon dioxide, argon being non-combustible gases. The composition of these gases is dependent upon the geographical region from where natural gas is extracted. The composition of natural gas may even vary from one gas well to another. The composition of natural gas is bound to affect the thermodynamic gas turbine cycle efficiency as molar composition of working fluid is flue gas (FG) varies after combustion in combustor, which in turn affects thermo-physical properties of working fluid, hence the need to evaluate the effect of combustible composition of natural gas on thermo-physical properties of working fluid circulating in the gas turbine. Table 1 shows the standard composition of natural gas as well as the composition range of individual components of natural gas extracted from a gas field.

The effects of natural gas composition on cycle performance are primarily due to:

Table 1 Composition of natural gas in particular gas field

Composition of natural gas	Weight percentage (%)
Methane (CH_4)	86.21
Ethane (C_2H_6)	7.20
Nitrogen (N_2)	5.56
Carbon dioxide (CO_2)	1.03

- Increase or decrease in heating value of natural gas due to variation in composition of CH_4 and C_2H_6.
- Effect of molar fraction of combustor flue gases (FG) on thermo-physical properties of flue gases mixture.

Survey of literature reveals reasonable work in the area of determination of thermo-physical properties of air and gas mixture. Spencer et al. [2] have provided empirical heat capacity equations of various gases for a temperature range of 298.1–1500 K. It supplies the values of thermodynamic functions of twenty-three substances at hypothetical ideal gaseous state. Cox et al. [3] have established thermo-physical properties of key chemical substances at 298.15 K. Lemmon et al. [4] have developed a model for the prediction of thermodynamic properties of gas mixtures. Annaratone [5] has provided a function for thermodynamic properties that have a good accuracy for $273 < T < 1673$ K with a percent error less than 1%, but error increases as the temperature increases above 1673 K. NASA has provided an equilibrium code for the determination of two temperature-dependent enthalpy polynomials, to cover a temperature range for most practical purposes, one polynomial for temperature $0 < T < 1000$ K and the other polynomial for $1000 < T < 5000$ K [6]. Lanzafame et al. [7–9] have proposed logarithmic polynomial of specific heat at constant pressure of gas species which is temperature-dependent and valid for the temperature range of 100–5000 K. The proposed logarithmic polynomial is likely to be more accurate as compared to other polynomial [4–6] and results obtained would be validated with the data on available NIST web-book and JANAF-Thermochemical-Tables [10], Chase [11], Gordon and McBride [12], Gurvich et al. [13, 14].

Literature survey related to thermodynamics of gas turbine cycle has been considered as well. A literature review of thermodynamics of gas turbine cycles reveals various research papers, hence only recent articles have been discussed here.

Thermodynamics of gas turbine cycle incorporating blade cooling has been reported by Sanjay et al. [15, 16]. The article compares gas turbine cycle with different blade cooling techniques. Most of the literature dealing with thermodynamics of gas turbine cycles have not adopted cycle having air cooled blades which is a more realistic approach for such analysis. Mahapatra et al. [18] proposed that increasing ambient temperature and ambient relative humidity of air leads to increment in blade coolant required for blade cooling.

Above literature survey identifies following research gaps as listed below and the same are addressed in this paper:

1. The mathematical model for determination of temperature and constituent dependent c_p of working fluid for gas turbine cycle analysis is sparse in literature.
2. Application of above developed thermo-physical property polynomial for gas turbine cycle analysis has not been reported.
3. Performance map of analyzed gas turbine cycle with the variation of composition of natural gas constituents has been wanting in literature.

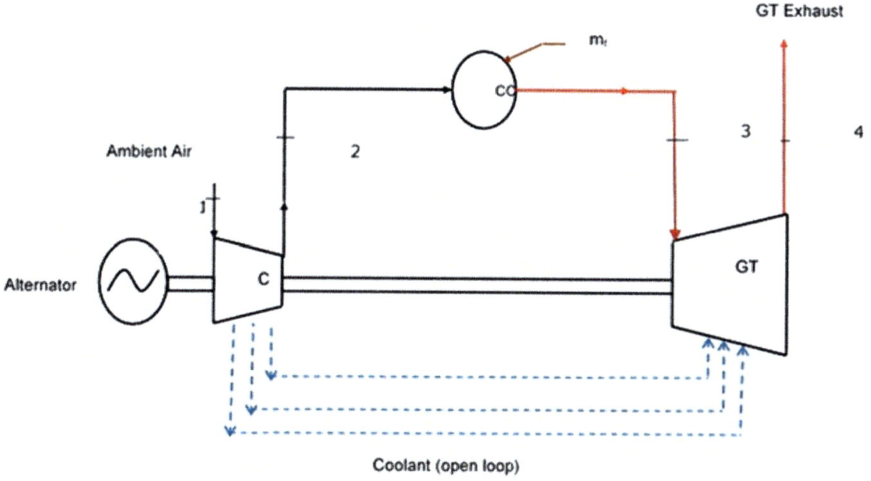

Fig. 1 Schematic view of gas turbine cycle

2 Gas Turbine Cycle Configuration

Figure 1 illustrates a schematic of a gas turbine subjected to film-air cooling of hot gas path components (blades). The proposed cycle comprises of standard gas turbine components, an axial flow multistage compressor, can-annulus DLN combustor, and a multistage expansion turbine. Air is ingested into the compressor at ambient condition of 288 K, 1.013 bar and gets compressed to a higher pressure leading to the rise in its temperature and hence enthalpy. Air for the cooling of gas turbine buckets is bled from air compressor at selected points and pressure level such that flow of coolant air to the blades takes place due to pressure differential. Compressed air exiting the compressor is routed to the combustor where natural gas is burnt giving rise to flue gas (FG) of varying composition depending upon the composition of combustible components of natural gas. High pressure, temperature gases exiting the combustor are expanded in the multistage expander section of the gas turbine.

3 Thermodynamic Modeling

Mathematical modeling of gas turbine cycle components has been done, and governing equations to predict cycle performance has been developed.

3.1 *Modeling of Cycle Components*

3.1.1 Combustion Chemistry

Natural gas comprises of various combustible and non-combustible gases, i.e., methane, ethane being the combustible components and nitrogen, carbon dioxide, argon being non-combustible gases. The composition of these gases is dependent upon the geographical region from where natural gas is extracted. The combustible components of natural gas being methane and ethane are burnt in the presence of excess air (150–300%) in the gas turbine combustor thus generating high temperature flue gases (FG). The combustion reaction for methane and ethane is given by:

Methane:

$$CH_4 + 2O_2 \rightarrow CO_2 + 2H_2O \tag{1}$$

Ethane:

$$2C_2H_6 + 7O_2 \rightarrow 4CO_2 + 6H_2O \tag{2}$$

3.1.2 Thermo-Physical Properties of Gas Turbine Flue Gases

The mass fraction of products of combustion and its temperature leads to dictate the thermo-physical properties of flue gases (FG) exiting the combustor. The specific heats of components of FG are as follows:

$$c_{p,N_2} = a_0 + a_1(\ln T) + a_2(\ln T)^2 + a_3(\ln T)^3 + a_4(\ln T)^4 + a_5(\ln T)^5 \tag{3}$$

$$c_{p,CO_2} = a_0 + a_1(\ln T) + a_2(\ln T)^2 + a_3(\ln T)^3 + a_4(\ln T)^4 + a_5(\ln T)^5 \tag{4}$$

$$c_{p,Ar} = a_0 + a_1(\ln T) + a_2(\ln T)^2 + a_3(\ln T)^3 + a_4(\ln T)^4 + a_5(\ln T)^5 \tag{5}$$

$$c_{p,H_2O} = a_0 + a_1(\ln T) + a_2(\ln T)^2 + a_3(\ln T)^3 + a_4(\ln T)^4 + a_5(\ln T)^5 \tag{6}$$

The c_p of mixture is calculated by using the formula:

$$c_p = c_{p,N_2} * X_{N_2} + c_{p,CO_2} * X_{CO_2} + c_{p,Ar} * X_{Ar} + c_{p,H_2O} * X_{H_2O} \tag{7}$$

3.1.3 Gas Model

Polynomial of specific heat of components of FG has been calculated through curve fitting based on data given by Lanzafame et al. [8] which is valid up to 5000 K.

$$c_{p,g} = a_0 + a_1(\ln T) + a_2(\ln T)^2 + a_3(\ln T)^3 + a_4(\ln T)^4 + a_5(\ln T)^5 \tag{8}$$

Enthalpy, entropy, and exergy of FG and air are calculated using polynomial (Eq. 8) in the equations here under:

$$h = \int_{T_a}^{T} c_p(T)\mathrm{d}T \tag{9}$$

$$\phi = \int_{T_a}^{T} c_p(T).\frac{\mathrm{d}T}{T} \tag{10}$$

$$s = \phi - R\ln\left(\frac{p}{p_a}\right) \tag{11}$$

$$\omega = h - T_a.s = h - T_a\phi + R.T_a \ln\left(\frac{p}{p_a}\right) \tag{12}$$

Here, all thermodynamic properties as enthalpy, entropy and availability associated with the components of (FG) are assumed to be zero at the ambient conditions, taken as 1.013 bar and 288 K and consider to be as datum level.

3.1.4 Compressor

An axial compressor is generally used in the power gas turbine. Thermo-fluid losses in an axial compressor have been in accounted for in the model by the adoption of suitable polytropic efficiency for the compressor. The temperature and pressure of air at any point in the compressor are given by the expression:

$$\frac{\mathrm{d}T}{T} = \left[\frac{R}{\eta_{pt,c} \cdot c_{p,a}}\right]\frac{\mathrm{d}p}{p} \tag{13}$$

Mass energy balance of the compressor is used to obtain compression work while the exergy losses are obtained from availability balance as per the following equations:

$$W_C = \dot{m}_{c,e} * h_{c,e} - \dot{m}_{c,i} * h_{c,i} \tag{14}$$

$$\Omega_{D,c} = W_C + \dot{m}_{c,i} * \omega_{c,i} - \dot{m}_{c,e} * \omega_{c,e} \tag{15}$$

Table 2 Data for analysis [15–17]

Parameter	Values/formulation	Unit
Gas-properties	$C_p = f(T)$	kJ/kg K
	Enthalpy $h = c_p(T)dt$	kJ/kg
Compressor	i. Polytropic efficiency (η_p) = 92.0	%
	ii. Mechanical efficiency (η_m) = 98.5	%
	iii. Ambient temperature of air = 288	K
	iv. Exit pressure = 0.5% of inlet pressure	bar
Combustor	i. Combustor efficiency (η_{comb}) = 99.5	%
	ii. Pressure loss (P_{loss}) = 2.5% entry pressure	bar
	iii. Fuel pressure = 1.5 ($p_{c,e}$)	bar
Gas turbine	i. Polytropic efficiency (η_{pt}) = 92.0	%
	ii. Turbine exhaust pressure = 1.08	bar
	iii. Turbine blade temperature (T_b) = 1123	K
Alternator	Efficiency ($\eta_{alternator}$) = 98.5	%

3.1.5 Combustor

Combustor-related losses are primarily due to pressure loss and incomplete combustion which are accounted for by the concept of combustion inefficiency (Table 2).

Mass of fuel required to achieve a specified combustor exit temperature (TIT) is determined using mass, energy, and exergy balance across the boundary of the combustor using equations as under:

$$\dot{m}_e = \dot{m}_i + \dot{m}_f \tag{16}$$

$$\dot{m}_f.\Delta H_r.\eta_{comb} = \dot{m}_e.h_e - \dot{m}_i.h_i \tag{17}$$

$$\Omega_{D,comb} = \dot{m}_f * \left[\Delta G_r + R_f * T_a * \ln\left(\frac{p_f}{p_a}\right)\right]_{comb} - \left[\dot{m}_e * \omega_e - \dot{m}_i * \omega_i\right]_{comb} \tag{18}$$

The combustor efficiency (η_{comb}) is taken to be 99.5%.

3.1.6 Air Film Blade Cooling Model

The gas turbine blades are subjected to open-loop air film cooling in which the process of mixing of coolant with the main gas stream leads to pressure drop and reduced fluid stream enthalpy. Air bled for cooling from compressor leads to drop in turbine output due to reduced mass of main gas flow expanding in the turbine. Further mixing of coolant air with main gas flow leads to drop in turbine exit temperature as well as pumping loss.

In air film cooling of blades, the coolant needs can be evaluated based on the modeling reported in author's previous work [15]. The mass flow rate of coolant to gas flow is given as:

$$a_{\text{coolant}} = \frac{\dot{m}_{\text{coolant}}}{\dot{m}_{\text{g}}} = \left[\frac{St_{\text{i}}.c_{\text{p,g}}}{\varepsilon.c_{\text{p,coolant}}}\right] \times \left[\frac{S_{\text{g}}.F_{\text{sa}}}{t\ .\cos\alpha}\right] \times \left[\frac{T_{\text{g,i}} - T_{\text{b}}}{T_{\text{b}} - T_{\text{coolant,i}}}\right] \tag{19}$$

$$R_{\text{c}} = \left[\frac{\left(T_{\text{g,i}} - T_{\text{b}}\right) \times c_{\text{p,g}}}{\left(T_{\text{b}} - T_{\text{coolant,i}}\right) \times \varepsilon.c_{\text{p,coolant}}}\right] \tag{20}$$

$$a_{\text{coolant}} = St_{\text{i}} \times \left(\frac{S_{\text{g}}.Fs_{\text{a}}}{t.\cos\alpha}\right) \times R_{\text{c}} = 0.0156.R_{\text{c}} \tag{21}$$

where $S_{\text{g}} \cong 2c$, $\frac{S_{\text{g}}}{t.\cos\alpha} = 3.0$, $F_{\text{sa}} = 1.05$, $St_{\text{in}} = 0.005$.

Also, the blade coolant requirement depends on temperature of coolant air at the bleed points, TIT, and blade material properties. Figure 2 details of the process of expansion in a cooled turbine stage.

Process b_1-c_1 illustrates cooling on account of heat transfer along constant pressure line between coolant and hot gas, on account of which enthalpy decreases, while process c_1-d_1 depicts mixing coolant with gas leading to drop in temperature (irreversible process) which leads to loss of entropy. d_1-a_2 represents process analogous to throttling in which enthalpy is conserved.

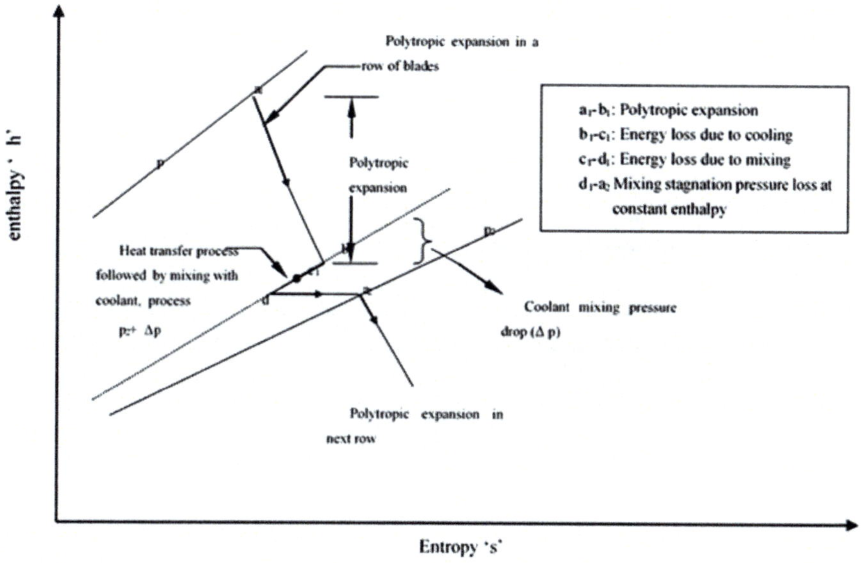

Fig. 2 Cooled gas turbine expansion process

3.1.7 Gas Turbine Expansion Mode

For evaluating gas turbine-specific work, the assumptions and equations used in the model are given as under:

1. The process of expansion in individual rows of gas turbine is taken to be polytropic which is accounted for in the model by assuming a suitable value of polytropic efficiency.
2. Every stage of turbine is assumed to be an expander, wall of which continuously extracts work from expanding gases.
3. The temperature at any expansion stage in the turbine is determined using

$$\frac{dT_g}{T_g} = \left(\frac{p + dp}{p}\right)^{\frac{R.\eta_{pt,t}}{c_{p,g}}} - 1 \tag{22}$$

4. The summation of work extracted by each row of turbine taken as and exergy destruction are given as:

$$w_{gt} = \sum \dot{m}_{g,i}.\left(h_{g,a_1} - h_{g,b_1}\right)_{\text{cooled}} + \sum \dot{m}_{g,i}.\left(h_{g,i} - h_{g,e}\right)_{\text{uncooled}} \tag{23}$$

$$\Omega_{D,gt} = \left(\dot{m}_{g,i} * \omega_{g,i} - \dot{m}_{g,e} * \omega_{g,e}\right) - W_{gt} \tag{24}$$

3.2 *Performance Parameters*

Expressions to determine the performance parameters of gas turbine cycle have been developed which include $W_{gt,net}$, η_{pl} and specific fuel consumption (sfc), exergy efficiency as under.

3.2.1 Net Gas Turbine Work ($W_{gt,net}$)

The network output of the turbine has been modeled to be dependent on the compression work and also dependent upon mechanical inefficiency (η_m) of the shaft train as under:

$$w_{gt,net} = w_{gt} - \frac{w_c}{\eta_m} \tag{25}$$

3.2.2 Plant Efficiency (η_{pl})

The cycle plant efficiency is given as:

$$\eta_{pl} = \frac{w_{gt,net}}{Q} = \frac{w_{gt,net}}{\dot{m}_f.\Delta H_r} \tag{26}$$

3.2.3 Specific Fuel Consumption (sfc)

Specific fuel consumption is expressed as fuel consumed in the combustor during one hour operation of gas turbine and is given as under:

$$\text{sfc} = \frac{3600.(\dot{m}_f/\dot{m}_a)}{w_{gt,net}} \tag{27}$$

3.2.4 Rational Efficiency (η_{rat})

Rational efficiency is taken as the ratio of actual electrical output to the input fuel exergy [17], and is expressed as:

$$\eta_{rat,gt} = \frac{W_{net,gt}}{\Omega_f} * 100 \tag{28}$$

4 Results and Discussion

Based on modeling discussed in previous Sect. 3, C++ coding has been done to compute results related to estimation of thermo-physical properties $c_{p,a}$,$c_{p,g}$, enthalpy, entropy as well as first and second laws of thermodynamic analysis of proposed gas turbine cycle when the combustible component (CH_4, C_2H_6) of natural gas is varied. Results obtained have been validated with work [19] and found to have an acceptable variance of around ±3%.

Figures 3, 4, 5, 6, 7, 8 and 9 show temperature-dependent variation of thermo-physical properties of various gases which constitute ambient air and gas turbine flue gases. The results obtained have been compared with thermo-physical dataset available from JANAF Thermochemical Table [10]. The proposed logarithmic polynomial is more accurate as compared to other polynomials [4–6], and results have been validated with JANAF Thermochemical Tables [10], Chase [11], Gordon and McBride [12], Gurvich et al. [13, 14], and error analysis of results shows variation between ±3% which is well within acceptable limit for

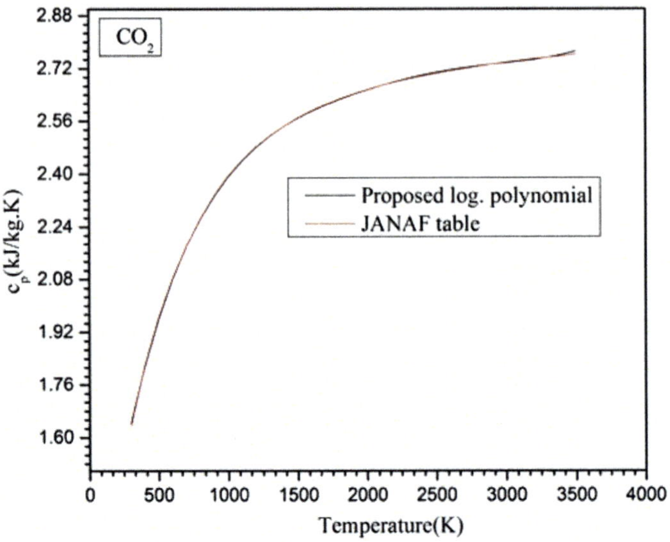

Fig. 3 Variation of specific heat of c_p, CO_2 at constant pressure versus temperature

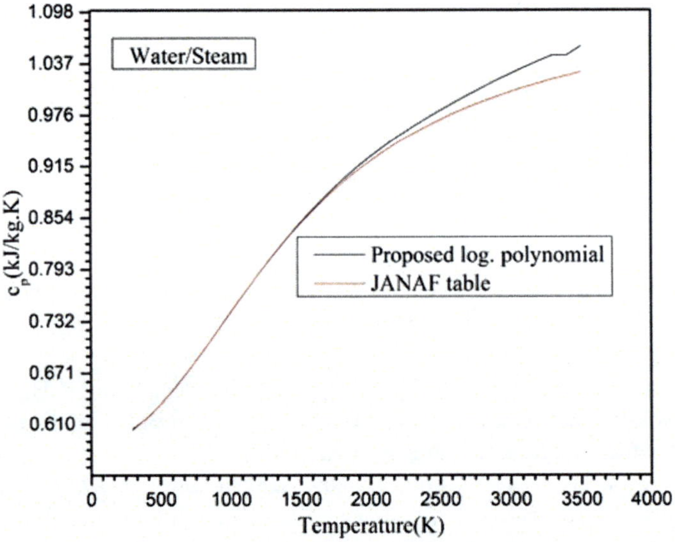

Fig. 4 Variation of specific heat of c_p, H_2O at constant pressure versus temperature

computational work. Hence, the developed fifth-order logarithmic polynomial has been adopted in thermodynamic analysis of the cycle results.

Figure 10 shows variation in gas turbine cycle net-specific work output of the proposed cycle with variation in TIT for two blends of natural gas having differing

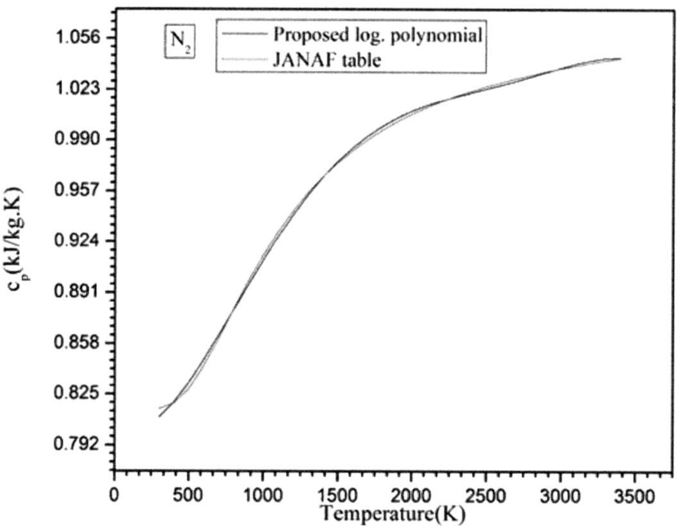

Fig. 5 Variation of specific heat of c_p, N_2 at constant pressure versus temperature

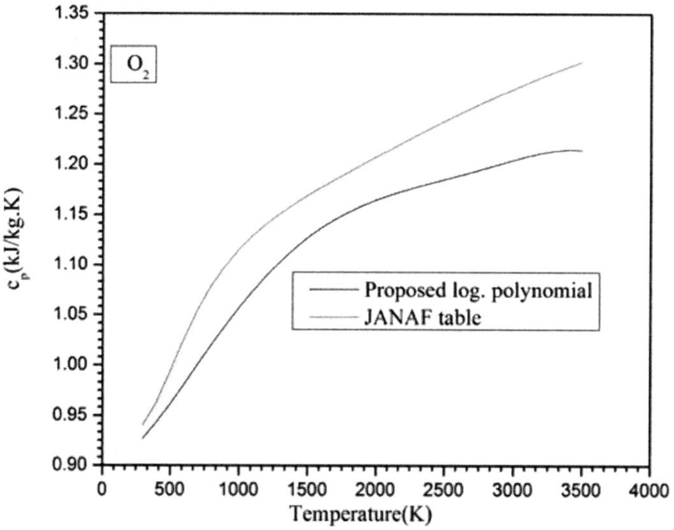

Fig. 6 Variation of specific heat of c_p, O_2 at constant pressure versus temperature

composition of methane and ethane. It is observed that gas turbine-specific work output increases with increase in TIT. This observed increase in cycle-specific work is primarily due to predominant increase in specific work on account of increase in TIT. Increase in TIT also leads to increase in coolant flow rate which increases

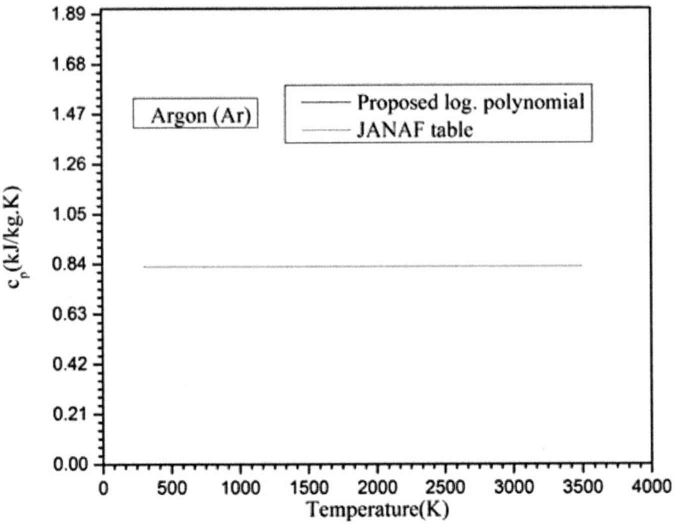

Fig. 7 Variation of specific heat of $c_{p,Ar}$ at constant pressure versus temperature

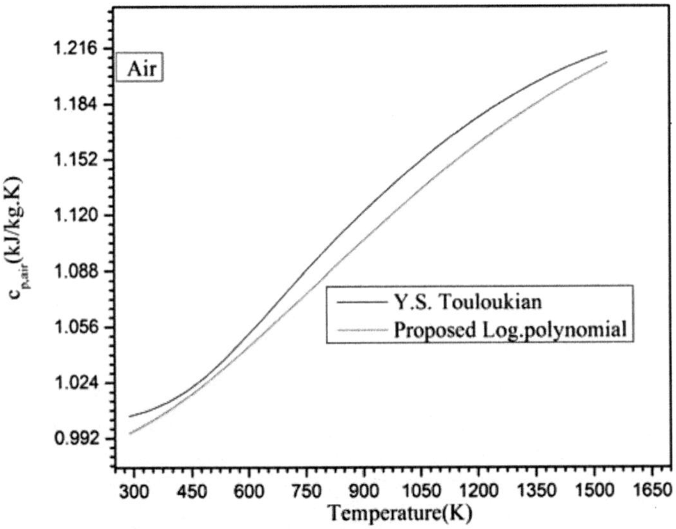

Fig. 8 Variation of specific heat of $c_{p,air}$ at constant pressure versus temperature

mixing, pumping, and cooling-related losses that reduce specific work but it is not so predominant.

Also, it is observed that the natural gas blend having higher CH_4 composition delivers superior gas turbine cycle net-specific work output, due to increased mass

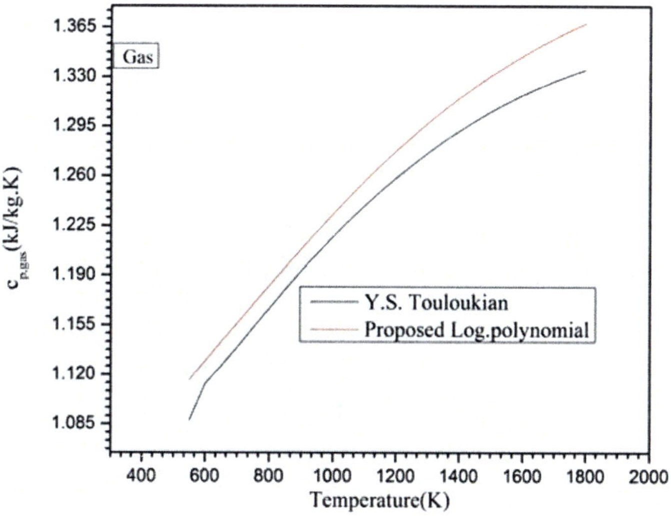

Fig. 9 Variation of specific heat of $c_{p.gas}$ at constant pressure versus temperature

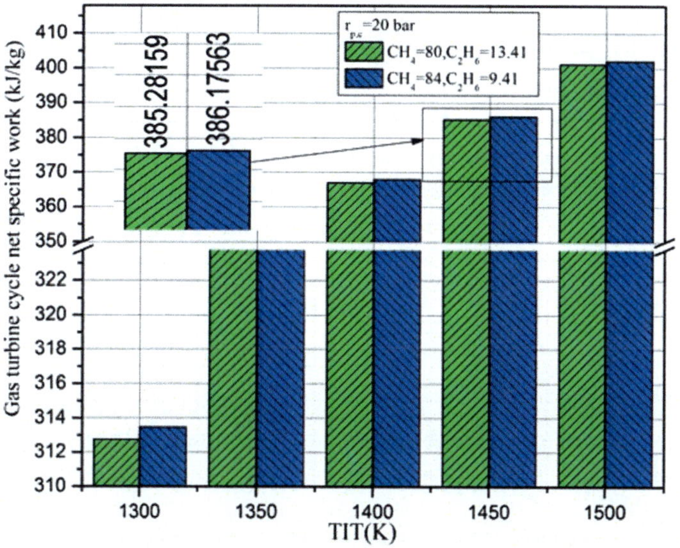

Fig. 10 Effect of variation in gas turbine (TIT) on gas turbine net-specific work

flow rate of fuel requirement which goes on to increase the mass flow rate of working fluid being handled by the cycle.

Figure 11 shows variation in gas turbine cycle net-specific work output of the proposed cycle with variation in compressor pressure ratio (r_{pc}) for two blends of

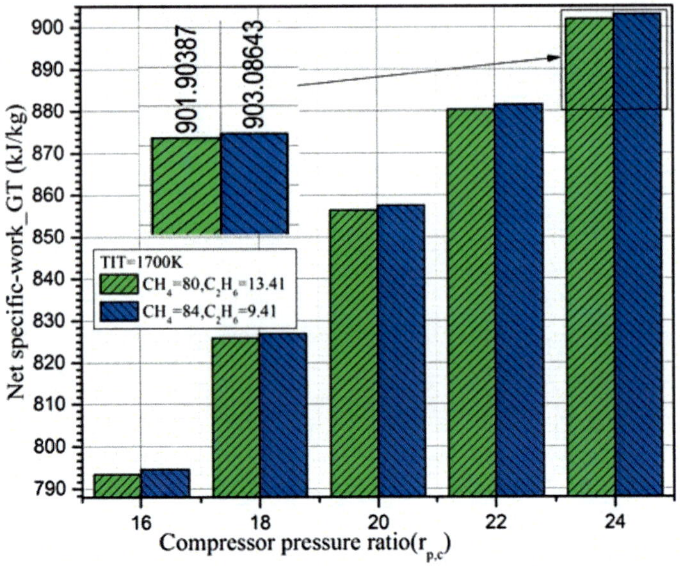

Fig. 11 Effect of variation in r_{pc} on gas turbine cycle net-specific work

natural gas having differing composition of methane and ethane. It is observed that gas turbine-specific work output increases with increase in r_{pc}. This observed increase in cycle-specific work follows the first law of thermodynamics. The observed increase in cycle-specific work primarily is due to predominant increase in specific work on account of increase in r_{pc}. Increase in r_{pc} also leads to increase in coolant flow rate which increases mixing, pumping, and cooling-related losses that reduce specific work but it is not so predominant.

Also, it is observed that the natural blend having higher CH_4 composition delivers superior gas turbine cycle net-specific work output due to increased mass flow rate of fuel requirement which goes on to increase the mass flow rate of working fluid being handled by the cycle.

Figure 12 shows variation in gas turbine cycle efficiency of the proposed cycle with variation in compressor pressure ratio (r_{pc}). It is observed that gas turbine cycle efficiency increases with increasing in r_{pc} and also with increasing composition of methane.

Figure 13 shows variation in gas turbine cycle efficiency of the proposed cycle with variation in TIT. It is observed that turbine efficiency increases with increasing in TIT and also with increasing composition of methane and then start decreasing after 1400 K. The reason associated with this behavior of graph is that gas turbine cycle efficiency increases with increasing TIT up to 1400 K because gas turbine cycle-specific work increases but after 1400 K coolant flow rates shoot up in the process significantly affecting cycle performances both in terms of thermal efficiency as well as specific work.

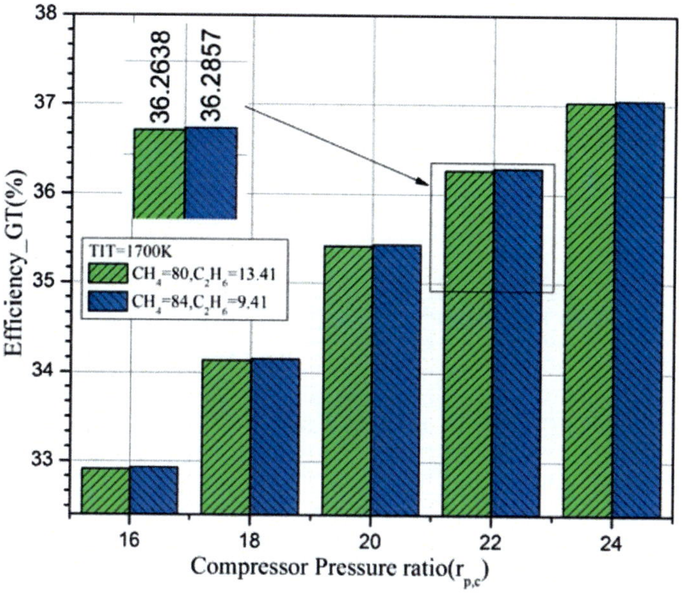

Fig. 12 Dependency on r_{pc} of gas turbine cycle efficiency at different natural gas composition

Fig. 13 Variation of TIT on gas turbine cycle efficiency at different natural gas compositions

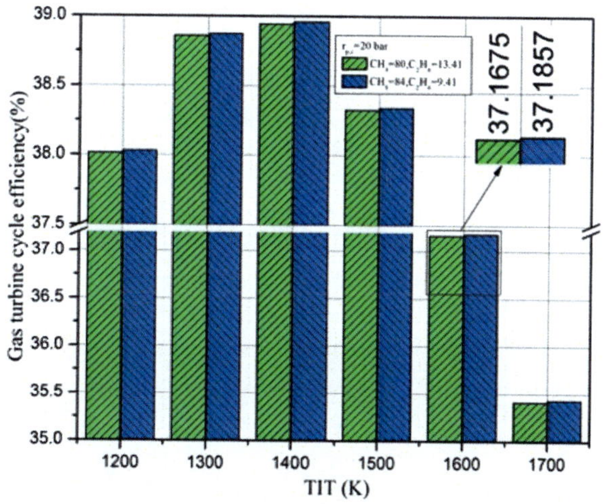

Figure 14 depicts the variation of turbine efficiency with specific work for various values of TIT and composition of methane. From performance map (Fig. 14), for a chosen value of methane composition and r_{pc}, the specific work increases with increase in TIT up to 1700 K whereafter it starts decreasing for TIT greater than 1700 K. This drop in performance of gas turbine at higher value of TIT is primarily due to increased blade coolant flow rates which bring down specific

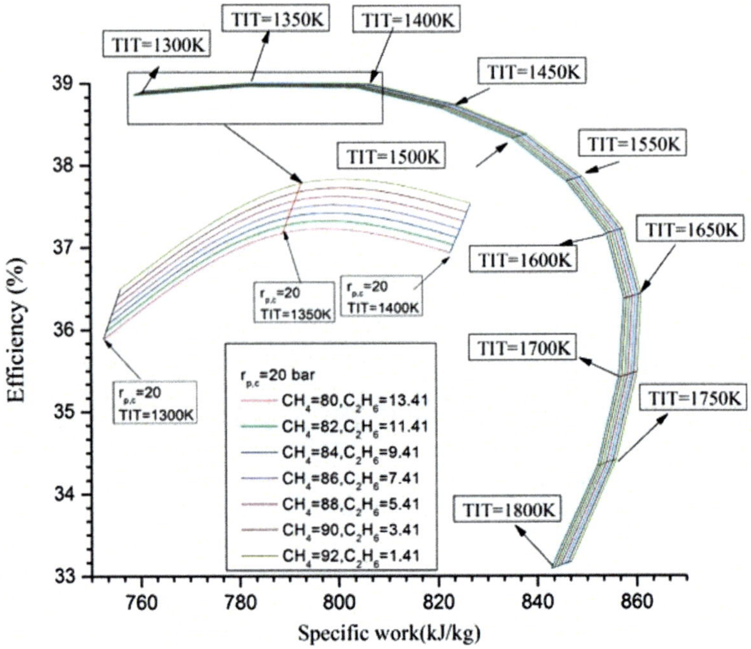

Fig. 14 Effect of turbine-specific work on efficiency at various natural gas compositions

power as well as thermal efficiency. Also, for a certain value of TIT, the cycle efficiency increases with increase in composition of methane.

The behavior of the trend shown in Fig. 14 can be explained as higher value of TIT leads to increase in mass flow rate of blade coolant which affects the gas turbine performance predominantly in specific work as well as in cycle efficiency.

Figure 15 shows a histogram of gas turbine cycle performance, illustrating variation in rational efficiency and exergy destruction (component-wise) at r_{pc} = 21, TIT = 1700 K with increasing composition of methane. Histogram shows that the rational efficiency decreases with increasing methane composition. However, maximum exergy destruction is in combustor followed by stack, gas turbine and compressor at a chosen value of composition of methane.

5 Conclusions

Based on the proposed mathematical model for temperature and molar mass fraction-dependent determination of thermo-physical property and thermodynamic analysis of gas turbine cycle, the following conclusions have been drawn:

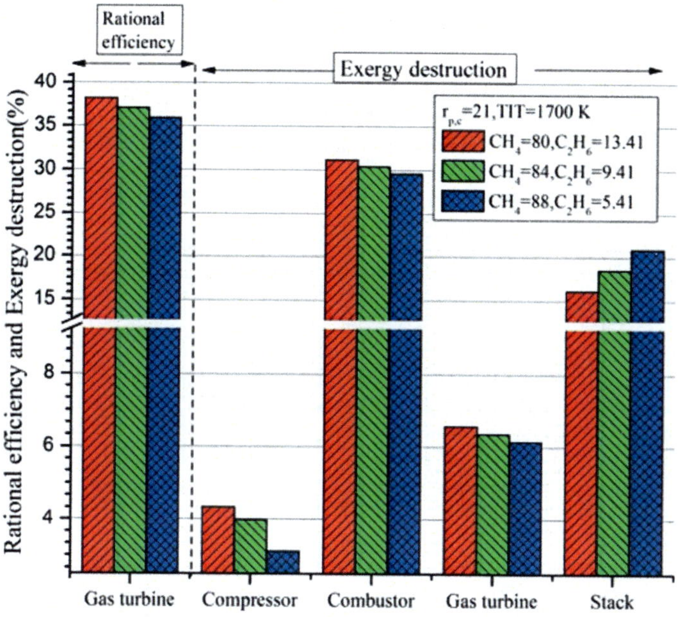

Fig. 15 Effect of rational efficiency and exergy destruction componentwise

- A fifth-order logarithmic polynomial for accurate determination of thermo-physical properties based on mole fraction of gas component has been developed.
- The thermo-physical properties of gas turbine working fluid based on the developed polynomial has been compared with NIST-JANAF and Y.S. Touloukain tables and found to be in good agreement.
- Composition of combustible gases in the natural gas fuel used in gas turbine is seen to effect thermo-physical properties of working fluid of gas turbine.
- Exergy analysis of gas turbine plant shows that combustion chamber exhibits highest exergy destruction of around 32%. Also, as percentage composition of methane increases, exergy destruction within combustion chamber decreases to 30% at 92% of CH_4.
- Plant rational efficiency is found to be 38% at CH_4 of 80% in the fuel.
- Hence, thermodynamic performance of gas turbine cycle is highest for natural gas fuel blend having higher percentage of methane composition.
- The summation of exergy destruction in all components is minimum ≈65% at higher percentage of CH_4 (92%) in the fuel.

References

1. International Energy Agency: World Energy Outlook Factsheet India. 9 rue de la Federation, 75739 Paris Cedex 15, France. www.iea.org (2015)
2. Hugh, M.: Empirical heat capacity equations of various gases. J. Am. Chem. Soc. **67**(10), 1859–1860 (1945)
3. Cox, J.D., Wagman, D.D., Medvedev, V.A.: CODATA key values for thermodynamics. Hemisphere Publishing Corp., New York. http://www.science.uwaterloo.ca/cchieh/cact/tools/thermodata.html (1984)
4. Lemmon, E.W., Jacobsen, R.T.: A generalized model for the thermodynamic properties of mixtures. Int. J. Thermo Phys. **3**, 825 (1999)
5. Annaratone Entalpia, D.: dei gas LaTermotecnica.-Settembre 1999-paper, pp 113–116 (1999)
6. Gordon, S., McBride, B.J.: Computer Program for the Calculation of Complex Chemical Equilibrium Composition, Rocket Performance, Incident and Reflected Shock and Chapman-Jouguet Detonation. NASA publication SP-273 (NTIS number N71–37775). dergipark.ulakbim.gov.tr/eoguijt/article/download/.../1034000159 (1971)
7. Lanzafame, R., Messina, M.: Thermodynamic property models for unburned mixtures and combustion gases. Int. J. Thermodyn. **9**(2), 73–80. http://dergipark.ulakbim.gov.tr/eoguijt/article/view/1034000172 (2006)
8. Lanzafame, R, Messina, M.: A new method for the calculation of gases enthalpy. In: Energy Conversion Engineering Conference and Exhibit (IECEC) 35th Intersociety, Las Vegas, NV, vol. 1, pp. 318–328 (2000)
9. Lanzafame, R., Messina, M.: Order Logarithmic Polynomials for Thermodynamic Calculations in ICE. SAE Technical Paper 2001-01-1912. https://doi.org/10.4271/2001-01-1912 (2001)
10. JANAF Thermochemical Tables 2nd ed. NSRDS-NB537, U.S. National Bureau of Standards (1971)
11. Chase, M.W.: NIST-JANAF thermochemical tables, fourth edition. J. Phys. Chem. Ref. Data Monograph **9**, 1-1951 (1998)
12. Gordon, S., McBride, B.: Thermodynamic Data to 20000 K for monoatomic gas. NASA/TP-208523 (1999). http://www.grc.nasa.gov/WWW/CEAWeb/TP-1999-208523.htm
13. Gurvich, L.V., et al.: Thermodynamic Properties of Individual Substances, vol. 1, issue 1 & 2. Nauka Moscow, URSS, and NASA Tech. Paper#. 3287 (1978)
14. Gurvich, L.V., Veyts, I.V., Alcock, C.B.: Thermodynamic Properties of Individual Substances, vol. 1, issues 1 & 2, 4th edn. Hemisphere Publishing Corp., New York and NASA Tech. Paper 3287 (1989)
15. Sanjay, Singh, O., Prasad, B.N.: Comparative evaluation of gas turbine power plant performance for different blade cooling means. Proc. IMechE Part A: J. Power Energ. **223**, 71–82 (2009)
16. Sanjay, Singh, O., Prasad, B.N.: Influence of different means of turbine blade cooling on the thermodynamic performance of combined cycle. Appl. Thermal Eng. **28**, 2315–2326 (2008)
17. Sanjay.: Investigation of effect of variation of cycle parameters on thermodynamic performance of gas-steam combined cycle. Energy **36**, 157–167 (2011)
18. Mahapatra, A.K., Sanjay.: Performance analysis of an air humidifier integrated gas turbine with film air cooling of turbine blade. J. Energy South. Afr. **24**(4), 71–81 (2013)
19. Touloukian, Y.S., Tadash, M.: Thermo-Physical Properties of Matter. The TPRC Data Series, vol. 6. IFI/PLENUNM, New York, Washington. www.dtic.mil/dtic/tr/fulltext/u2/a129114.pdf (1970)

NO*x* Elimination Laboratory Experiments for Biodiesel-Fueled C.I. Engine in Cooler Gas Flow Engine

Ravi Kumar Samadhiya and Devendra Kumar

1 Introduction

Without any fuel usage and/or engine performance penalties, biologic diesel emanates considerably fewer PM, carbon monoxide (CO) and hydrocarbon (HC) emissions. A particulate reduction effect can result in the lower aromatic and short paraffin HC chain and the higher oxygen content in biodiesel [1–3]. Carbon particles have been observed significantly smaller than the petrol engine in the cylinder head, the piston ring, and injectors of biodiesel fuel. Because of its inhuman lubricity properties, wear on critical rotating segments of the biodiesel engine is considerably less than the diesel engine [4–6]. Researchers have also shown that biodiesel NO*x* emissions have increased in comparison with diesel. And Salvatore et al. conducted direct turbo-loaded diesel injection experiments in rapeseed oil with methyl esters. It also said that the volume of NO*x* and decrease in Hc and CO and a sharp decline in smoke have been supported by methyl ester at the same time [7–9]. The exhaust gas temperature in jatropha biodiesel and its combined fuels increased with increased load and biodiesel. Ramesh et al. said biodiesel-powered motor CO_2 emissions are slightly higher than diesel fuel. "The biodiesel monoxide elimination was 8, 16 and 2.8 and 3.8 kW 16, 18, 20 and 22% higher biodiesel NO*x* emissions". The load conditions are of 2, 2.8, and 3.9 kW, respectively, for diesel emissions [10].

Exhaust gas recirculation is an effective method of NO*x* regulation. The exhaust gasses consist mostly of neutral carbon dioxide, heat, and nitrogen in a wide variety. Oxygen can be reduced and used as a heat sink when recirculated to the engine inlet [11]. This lowers the oxygen concentration and the temperature of the highest combustion, minimizing NO*x*. EGR is one of the most effective methods currently available to minimize NO*x* emissions from internal combustion engines.

R. K. Samadhiya (✉) · D. Kumar
Department of Mechanical Engineering, Sachdeva Institute of Technology, Farah, Mathura, UP, India

L. M. Das et al. (eds.), *Recent Trends in Thermal Engineering*,
Lecture Notes in Mechanical Engineering,
https://doi.org/10.1007/978-981-16-3428-4_18

However, the operation of EGR is still subject to sanctions. Unless correctly configured, smoke, fuel use, and heat output can be improved. EGR enables efficient regulation of greater NO*x* emissions.

Ming and Zheng et al. briefly studied the routes and the inconveniences to minimize NO*x* emissions from diesel and stressed the potential use of EGR. EGR has explored its effect on diesel operations and outlined a range of approaches to EGR inclusion [12]. The EGR flux and EGR hydrogen reform principles were proposed, too. New principles have also been suggested.

Biodiesel emissions are slightly lower with no fuel consumption penalties, or efficiency penalties for engine PM, carbon monoxide (CO), and hydrocarbon (HC). A lower aromatic and short paraffin HC chain and higher oxygen content could account for the effect of biodiesel on particulate reductions [13]. "Cylinder particles were found to be much smaller than diesel engine, the top of the piston, piston ring grooving, and the biodiesel injector". Because of its intrinsic lubricity properties, the wear on critical moving parts of the biodiesel engine are relative to the dogs. However, researchers have also shown higher emissions of NO*x* biodiesel than diesel.

That's Salvatore et al. have conducted tests on a direct injection turbo-loaded diesel engine using rape oil methyl esters. It was also stated that methyl ester increased NO*x* emissions, decreased HC and CO, and decreased considerably smoke during the same time as the injection [14]. For jatropha biodiesel and its mixed fuels, Ramesh et al. have stated that the "exhaust gas temperature" has been growing with an increase in freight and biodiesel volume. Contamination by biodiesel engines of CO_2 was substantially greater than diesel fuel. At 2, 2, 2, 2, 2, 2.5, and 3.5 kW, the decrease was 16, 14, and 14% of the carbon monoxide by biodiesel. Emissions of biodiesel NO*x* were 15, 18, and 19% above petrol. Emissions under load conditions were 2, 2, 2.5, and 3.5 kW, respectively.

A significant mechanism for NO*x* regulation is exhaust gas recirculation. A neutral carbon dioxide, nitrogen, and the high thermal value are the main components of the exhaled gas. It lowers the oxygen level and acts as a heat sink when recirculated to the motor inlet. This method decreases the content of oxygen and the peak temperature of the combustion, thereby minimizing NO*x*. EGR is one of the best ways in internal combustion engine to minimize NO*x* emissions usable at the moment. However, the EGR's actions need fines as well. Unless properly optimized, smoke, fuel consumption, and thermal performance can be improved significantly. Increased emissions of NO*x* are effectively tracked using EGR [15].

Al analysis outlines briefly the upcoming use of EGR for processes and deficiencies in the removal of diesel NO*x* emissions. There was a study of EGR's effect on diesel, and a range of approaches was highlighted for applying EGR. EGR stream therapy and EGR hydrogen reform were also proposed as new principles [16].

Biodiesel engines with improved exhaust gases were tested to produce less CO, hydrocarbons and mineral diesel contamination of particulate matter. The NO*x* diesel engine efficiently reduces exhaust gas (EGR), while the combustion chamber removes fuel temperatures and oxygen levels. Yet EGR contributes to higher particulate matter emissions (PM). Biodiesel EGRs have reduced NOx emissions, and no significant penalties have been applied to PM or BSEC contaminants. Including EGR decreases the NOx rate by a rise in EGR prices. The NOx reduction is higher at higher loads. The cause of the EGR production decrease in NOx is a decrease in oxygen volumes and a decreasing flame temperature.

The effect on the temperature and exhaust opacity of depleted gas recirculation, Agarwal et al. With the application of EGR, exhaust gas temperatures have fallen significantly. This ensures that NOx emissions can be minimized. The thermal performance and the easy use of brake petrol should not have a significant impact on EGR. However, the emissions of particulate matter rise during combustion, as can be seen with tests of smoke opacity.

2 Methodology and Experimental Setup

AF10 diesel engine was used primarily for farming practices and residential energy generation for current work on research: the selected Kirloskar builder, single cylinders, air cooled, direct injection. It is just the single four-stroke, vertical, air-cooled generator "normally suctioned". It has an electrical charge function, as it is combined with a flexible connection with such a phase inverter. The engine can be started with a decompression lever and has a centrifugal velocity control. The cylinder is made of cast iron and has a cast iron covered in rough phosphoric material. This engine's lubricating system is wet sump and the oil is filled with pumps on the front of a motor and is driven off the crankshaft on the pumping shaft. The input and the control valve are operated by a camship, which is supplied from either a crankshaft by two couples of gears. The pump is operated by a drive shaft end. Table 1 displays the functional motor details.

It is vital that the different instrumentation is mounted at the appropriate location in order to execute together the required test set and the necessary engine data. A recirculation system was also developed for the recirculation of the exhaust

Table 1 Technical configuration of the diesel engine

"Type and model"	Kirloskar DAF 10
"Category"	"Single cylinder, naturally aspirated, four strokes, vertical, air-cooled"
"Rated brake power (kW)"	7
"Rated speed (rpm)"	1560
Number of cylinder	One
Bore × stroke (mm)	102 × 110
Lubrication system	Forced feed
Compression ratio	17.5:1

portion of the motor inlet. A two-fuel tank system is required or vice versa for fast conversion between diesel and biodiesel. After completion of the data collection processes and collection of equipment, the data were placed on the display. Control panel of the MS Regulation has been developed, and on the front side of the control panel were devices like voltmeter, a meter, wattmeter, and rpm indicator. The electric charging bench with its controls on the front panel is on the rear of the control panel (12 bulbs per 580 W and 2 bulbs per 320 W). A 50-ml desk with stop cocks was also connected to the front of the panel to test diesel or organic diesel fuel flow. The two tanks with stop cocks on the back of a frame have been mounted at the highest point. Between the generator and the load table were a voltmeter, an ammeter, and a wattmeter. The motor body was fitted with a soiled nut and photo-reflection sensor. In order to calculate the exhaust temperature, the thermo-couples are mounted into the exhauster. For calculating various emission gas parameters, AVL 437 and AVL Di gas analyzer were kept nearby.

2.1 *Details of Experiment Setup*

See Fig. 1.

3 Result and Discussion

Initially, a series of motor tests was performed on a medium capacity diesel motor at 1500 min and at different rate EGRs to demonstrate the effect of EGR on smoke opacity and the amount of NOx exhaust. The EGR's effect on smoke and the exhaust quantity of NOx were initially performed by a series of engine tests at medium diesel engine power at 1,500 min and different EGR rates to measure the

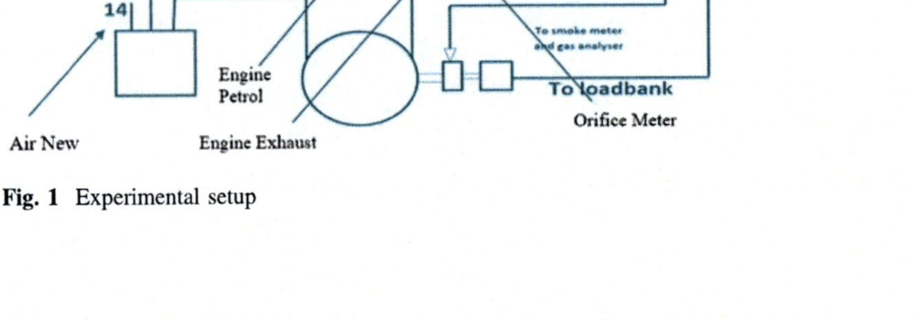

Fig. 1 Experimental setup

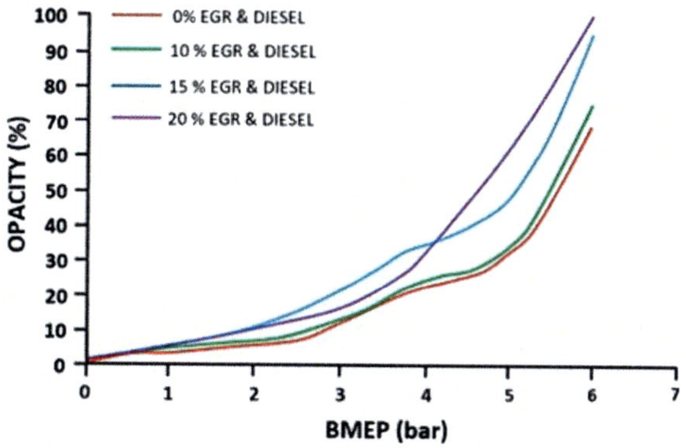

Fig. 2 BMEP opacity versus with numerous EGR ratios

PM in the exhaust gas; the exhaust gas smoke opacity is measured in Fig. 2, and Fig. 3 shows smoke opacity at varying levels of EGR. When EGR is used to run the motor, greater opacity to smoke than anything without EGR is observed. Smoke opacity increases with increased rates of EGR and increased engine load. The smoke opacity difference at high loads was higher than the lower. EGR decreases the supply of fuel oxygen, thereby leading to comparatively incomplete combustion and improving the output of PM. Figure 3 indicates the well-established value of EGR in reducing NO*x* diesel emissions. With the use of EGR, the rate of EGR decreases NO*x*. The NO*x* reduction is higher at higher loads. Low oxygen levels and low flame temperatures in diesel engines reduce NO*x* emissions by EGR.

EGR use reduces NO*x* in a diesel motor but increases opacity of the smoke. This is a famous NO*x*-PM exchange. On the other hand, the opacity of smoke is decreased and NO*x* is increased when biodiesel is used by diesel motors. EGR biodiesel may also be used to simultaneously reduce emissions of NO*x* and smoke. Detailed motor tests with 0–20% EGR and different concentrations of jatropha biodiesel, i.e., have been performed. B20, B50, and B100 are for the performance and emission pattern assessment of the motor. Data on production and emissions have been analyzed for thermal efficiency BSEC, HC, CO, NO*x* emissions and smoke opacity.

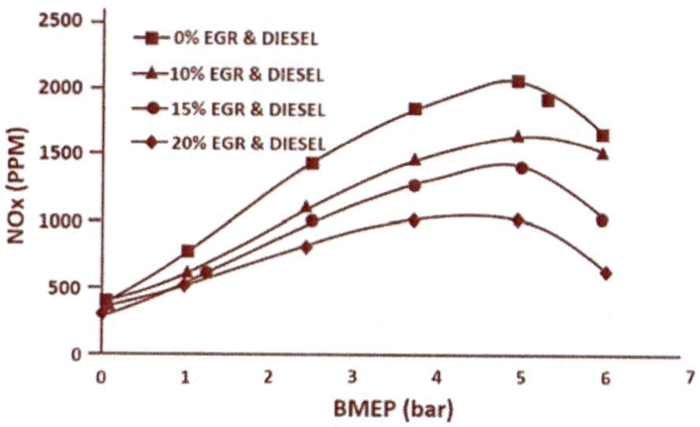

Fig. 3 BMEP NO*x* versus at various EGR speed

3.1 Performance Characteristics

The efficiency of the gasoline, blending and pure jatropha biodiesel test engine is summarized below.

3.1.1 EGR Performance

The test results showed that the thermal brake efficiency of all fuels was first improved as the braking power rose, followed by a further increase in brake capability. Trends in thermal efficiency are shown. Thermal efficiency is shown to be slightly improved with EGR at lower engine loads. The HCs which enter the combustion chamber can be reconstructed by means of the recirculated exhaust gasses. At higher engine loads, thermal efficiency remains the same as EGR.

When working in "biodiesel blends" with EGR with increasing biodiesel content in the mixture, thermal efficiency improved. The improvement in thermal efficiency of oxygenated fuels can be due to that. One of the interesting findings is the higher temperature efficiency of every biodiesel mix than the database lines shown in Fig. 4. For diesel at a load of 10% EGR of eighty percent, maximum heat output was observed. In B20, B50, and B100, the thermal peak was 27.95, 28.77, and 27.05%, while in diesel, the thermal peak was 26.79% and the EGR was 15%.

3.2 Brake-Specific Energy Usage

- Benefits of Brake Basic Fuel: The performance of two different fuels is not quite accurate because the density and heat value differ greatly. In order to compare

Fig. 4 Higher temperature efficiency of every biodiesel mix than the database lines

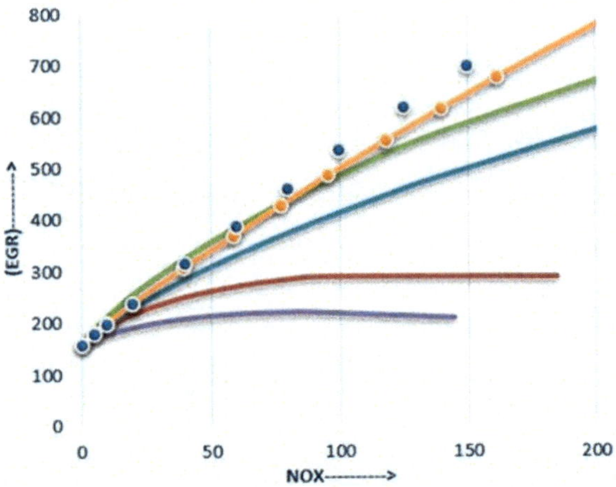

energy demand for unit power generation in different test fuels, the brake-specific energy was then taken as a metric.

- BSEC for diesel with EGR can be found to be less than baseline data at lower loads. But BSEC follows the same trend at increased loads with and without EGR. BSEC has been reduced with an increase in biodiesel concentration while the engine was running on biodiesel mixtures.
- The possible reason for lower BSEC may be improved thermal efficiency. The lowest BSEC was the 50% biodiesel blend of EGR. For tidy biological diesel with and without EGR, BSEC was better than base results at lower loads. In higher loadings, the engine has the same BSEC pattern in all datasets and BSEC has evolved at higher EGR speeds.

3.3 Emission Characteristics NOx Emissions

- The variations in NOx emissions for all study oils. As the engine load increased, the NOx emissions increased due to an increase in combustion temperature. This proves that the combustion temperature of the engine cylinder and the local stoichiometry of the mixture is a crucial factor in the emission of NOx.
- NOx contaminants were found at 80% load and 0% EGR at 1987 ppm for diesel and 2055 ppm for biodiesel in operation. Compared to petrol, the emissions of NOx for biodiesel fuel were higher. This may be due to higher biodiesel bulk modules which result in dynamic injection progress in addition to static injection progress, which ensures maximum efficiency. The condition may have been aggravated by excess oxygen (10%) in biodiesel. At higher loads, the volume of NOx was 5–8% higher than that of petrol. The degree of NOx reduction is higher at higher loads. Reduced levels of oxygen and flame temperatures are for the removal of EGR NOx emissions in diesel engines. However, in the case of biodiesel mixtures, NOx emissions are higher than gasoline due to higher temperatures in the combustion chamber.
- Here, it has been found that the lowest NOx emission of 15% EGR diesel is comparable to other study fuel samples. This is because inert gasses are present in the combustion chamber and molecular oxygen is not available. Twenty percent and more of EGR, the ratios could reduce NOx by significant amounts but reduced BTE and increased emissions of smoke, CO, and HC were observed.

3.4 Carbon Monoxide Emission

- Probably due to the dilution impact of exhaust gasses and the lower air–fuel ratio, the CO emissions can rise sharply at full load. The highest carbon dioxide emission was observed at 15% EGR for diesel and B20, which is likely to be due to low oxygen activity. Here, it is important to note that the combustion efficiency is faster at 15% and higher EGR speeds.
- The CO emissions of pure jatropha biodiesel fuel are lower than those of diesel fuel with no EGR over the entire load spectrum. This is possibly due to the abundance of molecular oxygen in biodiesel fuel. With an increase in EGR, CO emissions remain the same with lower loads but higher loads due to the dilution effect of the exhaust gas. In the case of biodiesel, 10% of EGR produced 1.04% (minimum) of CO emissions at full load and rapidly increased emissions to 22%.

3.5 Hydrocarbon Pollution Unburnt

- Higher concentrations of oxygen lead to low levels of air–fuel blends in the combustion chamber at different places. This heterogeneous mixture is not properly flammable and leads to higher HC emissions. The addition of biodiesel to diesel decreases the oxygen combustion needed due to the availability of molecular oxygen in gasoline. This leads to lower HC emissions; the emissions of HC are lower for biodiesel mixtures than diesel when the engine was operated by EGR. However, 18% of EGR pure biodiesel HC emissions are lower than baseline diesel emissions without EGR.
- Smoke opacity for EGR biodiesel mixtures is typically less than gasoline. Biodiesel molecule is made up of oxygen that is active in the combustion process, which can be used to contribute to better combustion and, eventually, to lower smoke. Pure biodiesel provides the lowest smoke opacity among all datasets of 15% EGR. Capacity increases without full load at increasing EGR speeds. When biodiesel is observed, there is a sudden rise in particulate matter at 20% EGR.

4 Conclusion

The test results show that biodiesel and its mixtures, EGR's full engine performances and emission characteristics are much higher than diesel fuel. When the motor was fueled with biodiesel and fired, when no gas exhaust was returned compared with diesel fuel, the thermal performance of the engine was lower and its use was higher for the brake. In the case of EGR, however, the requirements for biodiesel and its mixtures were superior. Without EGR, the emission of biodiesel NOx was higher than the emission of petrol. With rising EGR costs, the degree of NOx reduction was increased at higher loads. In the entire experimental area with and without use of EGR, carbon monoxide (CO), hydrocarbon (HC), and biodiesel smoke opacity were found to be smaller than gasoline. While NOx can be reduced by 20% or more, but BTE decreases and substantial increases in smoke, CO and HC emissions have been observed. The experimental results indicate that the ideal EGR strength effectively compensates for the opacity of NOx and smoke. From the experimental study, more than 15% of EGR values can be used at low loads, but at higher loads, the optimum EGR strength can be reduced to 15% to achieve better exhaust gas recirculation performance.

References

1. Agarwal, D., Sinha, S., Agarwal, A.K.: Experimental investigation of control of NOx emissions in biodiesel-fueled compression ignition engine. Renew. Energy **31**, 2356–2369 (2006)
2. Alfuso, S., Auriemma, M., Police, G., Prati, M.V.: The effect of methyl ester of rapeseed oil on combustion and emissions of D.I. diesel engines. SAE 932801 (1993)
3. Ramesh, D., Sampathrajan, A.: Investigations on performance and emission characteristics of diesel engine with Jatropha biodiesel and its blends. Agric Eng Int CIGR e-J Manuscript EE 07 013 **X** (2008)
4. Sharma, A.: An intelligent position control using priority-based fitness scheme and optimal tuning of fuzzy logic controller parameters with binary bat algorithm for nonlinear gantry crane system. J. Green Eng. **10**(11), 11687–11700 (2020)
5. Zheng, M., Reader, G.T., Hawley, J.G.: Diesel engine exhaust gas recirculation—a review on advanced and novel concepts. Energy Convers. Manage. **45**, 883–900 (2004)
6. Agrawal, A.K., Singh, S.K., Sinha, S., Shukla, M.K.: Effect of EGR on theexhaust gas temperature and exhaust opacity in compression ignition engines. Sadhana **29**(3), 275–284 (2004)
7. Mausam, K., Sharma, K., Bharadwaj, G., Singh, R.: Multi-objective optimization design of die-sinking electric discharge machine (EDM) machining parameter for CNT-reinforced carbon fibre nanocomposite using grey relational analysis. J. Braz. Soc. Mech. Sci. Eng. **41**(8) (2019)
8. Saraswat, M., Chauhan, N.R.: Comparative assessment of butanol and algae oil as alternate fuel for SI engines. Eng. Sci. Technol. Int. J. Elseveir. **23**(1), 92–100 (2020)
9. Saraswat, M., Sharma, K., Chauhan, N., Shukla, R.: Role of Automation in Energy Management and Distribution (online) Nopr.niscair.res.in (2020)
10. Sharma, P., Tiwari, M., Sharma, K.: Design and Analysis of a Process Plant Piping System (online). Inpressco.com (2020)
11. Shukla, M., Kumar, A., Yadav, A., Sharma, K.: Improved mechanical properties of graphene oxide reinforced cross-linked epoxy nanocomposites: a molecular dynamics approach. Mater. Today Proc. **11**, 679–685 (2019)
12. Saraswat, M., Sharma, K., Chauhan, N.R., Shukla, R.K.: Role of automation in energy management and distribution. JSIR **79**(10), 951–954 (ISSN0975-1084) (2020)

13. Verma, S., Sharma, K., Gupta, N., Soni, P., Upadhyay, N.: Performance comparison of innovative spiral shaped solar collector design with conventional flat plate solar collector. Energy **194**, 116853 (2020)
14. Sharma, A., Yadav, R., Chaturvedi, R.: Impact and analysis of axial vane installed in lift of Coanda UAV. Eur. J. Mol. Clin. Med. **7**(4), 708–718 (2020)
15. Yadav, A., Kumar, A., Sharma, K., Shukla, M.: Investigating the effects of amine functionalized graphene on the mechanical properties of epoxy nanocomposites. Mater. Today Proc. **11**, 837–842 (2019)
16. Saraswat, M., Chauhan, N.R.: Assessment study of butanol-gasoline blends in variable compression ratio spark ignition engine. JSIR **77**(7), 405–409 (ISSN0975-1084) (2018)